CISM International Centre for Mechanical Sciences

Courses and Lectures

Volume 576

The series presents lecture notes, monographs, edited works and proceedings in the field of Mechanics, Engineering, Computer Science and Applied Mathematics. Purpose of the series is to make known in the international scientific and technical community results obtained in some of the activities organized by CISM, the International Centre for Mechanical Sciences.

More information about this series at http://www.springer.com/series/76

Cristian Marchioli
Editor

Collective Dynamics of Particles

From Viscous to Turbulent Flows

Springer

Editor
Cristian Marchioli
Department of Engineering and Architecture
University of Udine
Udine
Italy

ISSN 0254-1971 ISSN 2309-3706 (electronic)
CISM International Centre for Mechanical Sciences
ISBN 978-3-319-84597-5 ISBN 978-3-319-51226-6 (eBook)
DOI 10.1007/978-3-319-51226-6

Printed on acid-free paper

This Springer imprint is published by Springer Nature
The registered company is Springer International Publishing AG
The registered company address is: Gewerbestrasse 11, 6330 Cham, Switzerland

Preface

Particulate flows are present in many natural and industrial processes. Transport of sediment in rivers and estuaries, convection of pollutants in the atmosphere, bio-convection of zooplankton, gravity, and turbidity currents near coastal shore, and pyroclastic flows from volcanic eruptions are a few examples that can be encountered in natural phenomena. In industry, processes involving flows of particles are numerous: among others, fluidized bed reactors, the treatment of waste materials in clarifiers, food processing, and ink technologies. In all the above-mentioned instances, proper understanding and accurate modeling of such complex flows are crucial aspects from scientific and engineering perspectives, as they directly impact the environment we live in. The understanding of such flows is a daunting task for several reasons. The most straightforward is the very large number of particles one needs to account for. Another equally significant difficulty arises from the subtle coupling between particle–particle and particle–fluid interactions: Particles have an effect on the fluid flow (and sometimes even drive it) by exerting stresses on the fluid around them, and in turn, the fluid flow modifies the motion of the suspended particles. This two-way coupling often makes attempts at comprehending such flows highly difficult, other than in very simplified settings. Particulate flows have been examined in the past in a wide variety of situations. A very large number of studies have focused on highly viscous flows in which inertial forces can be neglected. This low Reynolds number limit is a valid approximation in small-scale systems or very slow flows, and is often justified when the size of the particles involved in the process is small. In many practical applications, however, fluid inertia cannot be neglected owing to the large system sizes, even when the suspended particles are small. In some cases, it is of fundamental importance such as in pyroclastic flows or in fluidized bed reactors where the flows are highly turbulent in spite of the microscopic size of the particles involved. Several studies have focused on the Lagrangian properties of particles in turbulence (e.g., Lagrangian acceleration) to gain further insight on the relevant forces acting on isolated particles. Preferential concentration and clustering effect of inertial particles in a turbulent flow have been also examined in many recent works. However, collective effects in turbulent particle-laden flows have been not thoroughly examined, and there is a compelling need to provide a robust body of knowledge in this active field of research. The scope of this book is therefore to provide a state-of the-art and accessible survey of numerical approaches as well as modeling tools for the analysis of collective dynamics of particles in flows. The general approach is made specific through the most tractable analytically case of low Reynolds flows but goes beyond viscous flows and tackles inertial and turbulent flows. This book also covers the two main avenues for addressing particulate flows: one being discrete particle simulations and the other being continuum two-phase modeling. In the later, the influence of particles is captured through constitutive relations often resulting from simulations or experiments. The most common discrete methods for the description of particle-laden flows, both in the Stokes regime and in the inertial and turbulent regimes, are presented and discussed. Among the topics included are finite-size particles, and particles of different shapes, in particular rod-like particles or fibers whose interest lies in part in the availability of methods for slender bodies as well as in their importance in industrial applications, such as the fabrication of fiber-reinforced materials and of pulp and paper. This book provides a comprehensive overview of particulate flows, from low Reynolds numbers to full turbulent flows and hence can be particularly attractive to graduate students, Ph.D. candidates, young researchers, and faculty members in applied physics, chemical engineering, and mechanical engineering. The advanced topics and the presentation of current progress in this very active field may also be of considerable interest to many senior researchers, as well as to industrial practitioners having a strong interest in understanding the multiscale complex behavior of such multiphase flows.

This book contains selected printouts based on the lectures given during the advanced course entitled "Collective Dynamics of Particles: From Viscous to Turbulent Flows." The course was held at the International Center of Mechanical Sciences in Udine (Italy), organized under the auspices of ERCOFTAC and with the support of COST, through Action FP1005 "Fiber suspension flow modeling" and ANR CoDSPiT "Collective dynamics of settling particles in turbulence." The contributions cover introductory concepts related to modeling and simulation of spherical particles (Maxey and Dent, Bourgoin, Homann) and rod-like particles (Butler) in fluid flow, and collective effects at low (Maxey and Dent, Bourgoin) and finite particle Reynolds number (Maxey and Dent, Homann, Butler) as well as low-flow (Maxey and Dent, Butler) and high-flow Reynolds numbers (Maxey and Dent, Bourgoin, Homann).

The authors wish to thank all the contributors and the members of the International Center of Mechanical Sciences, in particular Prof. Elisabeth Guazzelli, for her continuous support and thoughtful suggestions during the preparation of this volume.

Udine, Italy Cristian Marchioli

Contents

Modeling and Simulation of Discrete Particles in Fluid Flow

Martin R. Maxey and Gelonia L. Dent

Abstract A summary is given of some of the methods for modeling and simulating the motion of small rigid particles in fluid flow. For isolated particles, or at very low volume fractions, approximate dynamic equations for tracking the motion of the particles can be formulated where the background flow is not modified. Even here, interesting features of the particle motion and distribution of particles can develop in nonuniform flows. For more complex situations, we describe the force coupling method (FCM) as an effective representation for particles moving at low and finite Reynolds numbers that may be applied to various dispersed multiphase flows.

1 Introduction

The study of particle motion in fluid flows relates to many different areas. In the atmosphere, small water droplets may form by condensation from vapor in the surrounding air and then continue to grow by sporadic collisions with other droplets, possibly forming precipitation (Shaw 2003; Grabowski and Wang 2013). The droplets range in size from a few microns to a millimeter or so and are transported by the turbulent air flow while settling slowly under gravity. An atmospheric cloud of droplets may appear to be opaque but in fact the droplets are quite disperse with a volume fraction substantially less than 0.1%. The fluid (air) forces on the droplets create equal and opposite forces on the air flow and if the mass loading, i.e. the relative mass fraction of droplets in a given volume, is appreciable there can be a significant negative buoyancy on the flow. Since the density of water is about a

M.R. Maxey (✉) · G.L. Dent
Division of Applied Mathematics, Brown University, Providence, RI, USA
e-mail: Martin_Maxey@Brown.edu
URL: https://www.brown.edu/academics/applied-mathematics/martin-maxey

G.L. Dent
Medgar Evers College, City University of New York, Brooklyn, NY, USA
e-mail: gelonia_dent@brown.edu
URL: http://www.brown.edu/academics/applied-mathematics/gelonia-dent

C. Marchioli (ed.), *Collective Dynamics of Particles*,
CISM International Centre for Mechanical Sciences 576,
DOI 10.1007/978-3-319-51226-6_1

thousand times that of air this can easily occur and leads to a sinking motion of the cloud and the surrounding air mass.

There is a general classification of dispersed two-phase, particle-laden flows based on the relative volume fraction and mass loading, see Balachandar and Eaton (2010) and references therein. The combination of both low volume fraction and low mass loading gives one-way coupling. The overall flow is not modified by the particles, they essentially move in isolation, and they respond to fluid forces, such as drag, generated by their motion relative to the local ambient flow. Weak two-way coupling occurs if the overall flow is modified by the particles due to the mass loading but they are still sufficiently dispersed that the volume fraction remains low and they move in isolation with little fluid dynamic interaction between them. Full two-way coupling describes the general context of significant mass loading and volume fraction. One must then consider the particles and fluid as a collective system as in a suspension flow or a slurry flow, commonly involving solids in liquids. There are further classifications based on the size of the particle relative to the scale of the fluid motion and the Reynolds number associated with the particle motion.

There are numerous references covering the motion of particles in a range of interesting contexts. Guazzelli and Morris (2012) provides a good overview of suspension flows for low Reynolds number conditions. Di Carlo (2009) reviews techniques for separating small particles or cells by size for biomedical applications relying on the role of inertial forces. Meiburg and Kneller (2010) give a review of turbidity currents and flows involving sediment transport in estuaries or coastal waters.

In this chapter, we first describe some of the ways to quantify the motion of isolated rigid particles considering mostly smaller spherical particles and the range of fluid forces that should be considered. We then consider the motion of these particles in simple laminar flows to illustrate how discrete particles differ from Lagrangian fluid tracers and some of the phenomena that may arise in nonuniform flows. In the subsequent sections we describe the force coupling method as a way to explore particle motion in more complex situations, including the effects of fluid interactions between particles and applications to suspension flows.

2 Motion of Isolated Particles

In the simplest situation a particle is so small that the viscous fluid forces, which scale with the surface area of the particle, are strong enough that there is no relative motion between the particle and the surrounding fluid. The particle then moves as a Lagrangian fluid element. The position $\mathbf{X}(t)$ of such a fluid element in an incompressible flow $\mathbf{u}(\mathbf{x}, t)$ is then given by

$$\frac{d\mathbf{X}}{dt} = \mathbf{u}(\mathbf{X}(t), t) \tag{1}$$

This concept is the basis of particle image velocimetry (PIV) (Adrian and Westerweel 2011), where the flow is seeded with a sufficient number of small tracer particles to resolve the flow field but still maintaining a very low volume fraction and mass fraction. We assume that the effects of Brownian motion are negligible relative to the fluid forces and motion. The displacements of the PIV particles over short time intervals may then be used to measure the local fluid velocity. Further, these small spherical particles will rotate at an angular velocity $\boldsymbol{\Omega}$ equal to half the local fluid vorticity $\boldsymbol{\omega}(\mathbf{X}(t), t)$.

2.1 *Gas–Solid Flows*

The density of a solid particle is usually much larger than that of a gas and the effects of gravitational settling or particle inertia may be significant even if the particle is very small, essentially a point, relative to the scale of the surrounding flow. In the absence of a flow, a small but heavy particle will eventually fall at a terminal velocity $\mathbf{W}_S$ given by a balance of the fluid drag force in steady motion and the force of gravity. Buoyancy effects are negligible and using a Stokes drag law for a sphere of radius a (Batchelor 1967) the fluid force on a particle of mass m_P is,

$$\mathbf{W}_S = m_P\mathbf{g}/6\pi a\mu$$

where $\mathbf{g}$ is the acceleration due to gravity and μ is the dynamic viscosity. A corresponding estimate may be made when the particle Reynolds number $\mathrm{Re}_P = 2aW_S/\nu$ is non-zero and where ν is the kinematic viscosity of the fluid.

More generally, we should consider the inertia of the particle and the approximate equation of motion for a small particle, where $\mathrm{Re}_P \ll 1$, is

$$m_P\frac{d\mathbf{V}}{dt} = 6\pi a\mu\,(\mathbf{u}(\mathbf{Y}, t) - \mathbf{V}) + m_P\mathbf{g}. \tag{2}$$

This gives the particle velocity $\mathbf{V}(t)$ and position $\mathbf{Y}(t)$ in terms of the force of gravity and the Stokes drag force based on the instantaneous velocity of the particle relative to the local fluid flow. The key assumption is that the fluid drag force comes quickly to a quasi-steady equilibrium. This may be rewritten as

$$\tau_P\frac{d\mathbf{V}}{dt} = (\mathbf{u}(\mathbf{Y}, t) - \mathbf{V}) + \mathbf{W}_S, \tag{3}$$

where τ_P is the inertial response time of the particle $\tau_P = m_P/6\pi a\mu$. So a particle falling from rest will eventually reach the terminal fall speed on the time scale τ_P and similarly the particle velocity will adjust to changes in the local flow velocity on the same time scale. Either way τ_P is taken to be longer than the time for the fluid drag force to become fully developed and reach its quasi-steady value. An important parameter is the Stokes number $St = \tau_P/T_F$, the ratio of the inertial response time to the time scale of variation in the local ambient flow velocity $\mathbf{u}(\mathbf{Y}(t), t)$. In the limiting case of $St \ll 1$, the particle velocity is the sum of the local flow velocity and the fall velocity,

$$\mathbf{V} = \mathbf{u}(\mathbf{Y}, t) + \mathbf{W}_S. \tag{4}$$

These equations describe the motion of solid aerosol particles or water droplets under many conditions. The dynamic viscosity of water is some 50–60 times that of air and a small droplet, held spherical in shape by surface tension, will largely respond as a rigid particle. A droplet of radius $a = 10\,\mu\text{m}$ will fall under standard conditions at about $W_S = 1.2\,\text{cm}\,\text{s}^{-1}$ and $\text{Re}_P = 0.016$. The Stokes drag law is a good estimate for the fluid force. The particle Reynolds number grows as a^3 in this range, but even for a droplet of radius $a \sim 25\,\mu\text{m}$ the error in the Stokes estimate of the drag force is less than 5% and $\text{Re}_P \sim 0.2$.

2.2 *Liquid–Solid Flows*

When gravitational settling or particle inertia is relevant in other contexts, such as solids in liquid flows, one needs to consider more fully the motion of a particle, even if it is isolated. A sphere moving relative to the surrounding fluid will create a local disturbance flow, which gives rise to fluid forces such as viscous drag or an added-mass effect. If there is an ambient flow $\mathbf{u}(\mathbf{x}(t), t)$ without any particle, this is locally modified to become an incompressible flow $\mathbf{v}(\mathbf{x}(t), t)$ that satisfies

$$\left(\frac{\partial \mathbf{v}}{\partial t} + \mathbf{v} \cdot \nabla \mathbf{v}\right) = \mathbf{g} - \rho^{-1} \nabla p + \nu \nabla^2 \mathbf{v}. \tag{5}$$

The boundary conditions are that $\mathbf{v} = \mathbf{V} + \boldsymbol{\Omega} \times [\mathbf{x} - \mathbf{Y}(t)]$ on the sphere, at $|\mathbf{x} - \mathbf{Y}(t)| = \text{a}$, and $\mathbf{v} = \mathbf{u}(\mathbf{x}, t)$ as $|\mathbf{x} - \mathbf{Y}(t)| \to \infty$. The flow configuration is illustrated in Fig. 1.

This problem can be solved approximately if the particle Reynolds number Re_P is small and the shear Reynolds number $\text{Re}_\Gamma = a^2\Gamma/\nu$ is small, where $\Gamma \sim U/L$ is representative of the velocity gradient in the ambient flow. The key steps involved are as follows.

- Change to a frame of reference moving with the particle: $\mathbf{z} = \mathbf{x} - \mathbf{Y}(t)$ and $\mathbf{w} = \mathbf{v} - \mathbf{V}(t)$

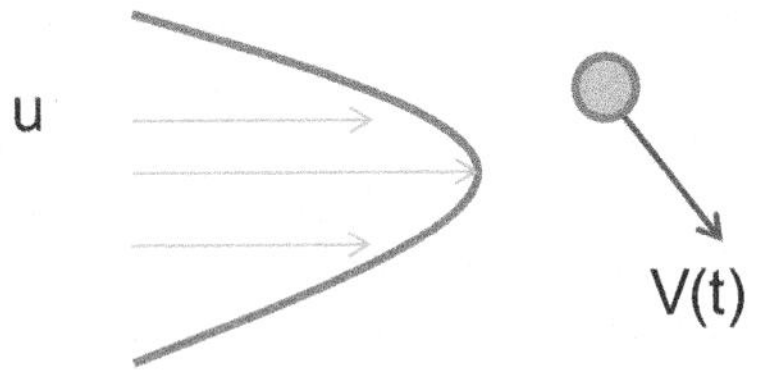

Fig. 1 Sketch of a particle located at $\mathbf{Y}(t)$ in an ambient flow $\mathbf{u}(\mathbf{x}, t)$

- Split $\mathbf{w}$ into the base flow $\mathbf{w}^{(0)} = \mathbf{u} - \mathbf{V}$ and a local disturbance flow $\mathbf{w}^{(1)}$
- Solve for $\mathbf{w}^{(1)}$ as an unsteady Stokes flow, where $\partial \mathbf{w}^{(1)}/\partial t$ is retained but $\mathbf{w}^{(1)} \cdot \nabla \mathbf{w}^{(1)}$ is neglected. This is most easily done using Laplace transforms for the time dependence.
- Evaluate forces due to the base flow
- Evaluate forces from the disturbance flow using a generalized form of the Reciprocal Theorem, adapted for unsteady Stokes flow. This involves integration of flow variables on the particle surface or over the particle volume.

These steps are summarized in Maxey and Riley (1983), together with references to earlier work. Equivalent results are also given by Gatignol (1983). Basset (1888) developed the original result for the unsteady motion of a sphere in viscous flow without any background flow.

The resulting equation of motion for the particle is

$$m_P \frac{d\mathbf{V}}{dt} = (m_P - m_F)\mathbf{g} + m_F \frac{D\mathbf{u}}{Dt} - \frac{m_F}{2}\frac{d}{dt}\left\{\mathbf{V} - \mathbf{u} - \frac{1}{10}a^2\nabla^2\mathbf{u}\right\}$$
$$- 6\pi a\mu \left\{\mathbf{Q} + a\int_0^t \frac{d\mathbf{Q}}{d\tau}[\pi\nu(t-\tau)]^{-1/2}d\tau\right\} \tag{6}$$

The mass of fluid displaced by the particle is m_F while the term $\mathbf{Q}(t)$ is defined as,

$$\mathbf{Q} = \mathbf{V}(t) - \mathbf{u}(\mathbf{Y}(t), t) - \frac{1}{6}a^2\nabla^2\mathbf{u} \tag{7}$$

For steady state motion in a uniform flow (6) reduces to a simple balance between the resultant force due to gravity and the viscous Stokes drag force. In a nonuniform flow, the Stokes drag law is not modified by the presence of a simple uniform shear flow but there is a Faxen correction for the finite size of the particle in a nonuniform flow with a quadratic variation as may arise in a Poiseuille flow, as given by (7). The equation (6) assumes as an initial condition that $\mathbf{Q}(0) = 0$.

There are several terms on the right hand side of (6) that balance the particle inertia on the left and represent different aspects of the flow response to the particle motion. Beyond the effect of gravity or buoyancy, the next term gives the fluid force that the ambient flow would exert on an equivalent mass of fluid and equals the resultant force from pressure and viscous stresses in the ambient flow. The term after this is the effect of added mass, with a possible correction for the finite size of the particle. The last term represents the viscous forces.

Figure 2 illustrates the response to a particle accelerating from rest at a constant rate $\mathbf{A}$ in the absence of an ambient flow, so that $\mathbf{V} = \mathbf{A}t$. The first effect is an essentially inviscid pressure response due to the acceleration of the surrounding fluid about the particle. The initial primary balance of forces is between the added mass effect, particle inertia and the applied external force. The last term, involving $\mathbf{Q}$, represents the effect of viscous forces and the vorticity layer $\delta(t)$ generated by the

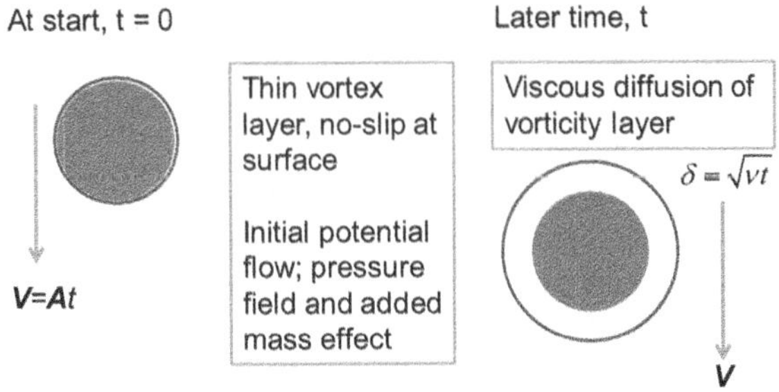

Fig. 2 Development of the disturbance flow $\mathbf{w}^{(1)}$ for a particle accelerating from rest at constant rate $\mathbf{A}$ with no ambient flow. The initial inviscid response of inertia and added-mass is followed by a developing vorticity layer δ that diffuses by viscous action

no-slip conditions on the particle as it gradually accelerates. The layer grows in scale by viscous diffusion, $\delta \sim \sqrt{\nu t}$. The second part of this last term is usually referred to as the Basset history or integral term and captures the initially strong viscous shear stress generated by the vorticity layer. Eventually, the viscous force relaxes to the equilibrium Stokes value over a time scale of $O(a^2/\nu)$. The external force required to sustain this motion is $\mathbf{F}^{ext}$, given by

$$\mathbf{F}^{ext} = \left(m_P + \tfrac{1}{2}m_F\right)\mathbf{A} + 6\pi\mu a\mathbf{A}t\left\{1 + 2a(\pi\nu t)^{-1/2}\right\} \tag{8}$$

From (8) it is evident that while the vorticity layer is very thin at first, the velocity $\mathbf{V}$ is also small and the initial viscous force is small. Basset (1910) gives a corresponding result for the motion of a sphere falling from rest under gravity. The result is more complicated but has similar features and the final adjustment to the Stokes terminal velocity essentially decays as $a(\nu t)^{-1/2}$.

In a gas–solid flow, $m_P \gg m_F$, and so the inertial response time $\tau_P \gg a^2/\nu$. This means that the viscous drag force on the small particle quickly adjusts to the Stokes value as compared to the time scale τ_P on which the particle velocity is changing. Generally the effects of added mass and the force of the ambient flow are negligible too and so (6) reduces back to (2). For a solid particle in a liquid, the values of m_P and m_F are more similar and these additional effects are more relevant. Indeed for a small bubble, that responds as a rigid spherical inclusion, the added mass is the only inertial term and history effects can be significant (Candelier et al. 2004).

Another way to view these different contributions of the local disturbance flow to the fluid force on a particle is to consider the forces needed to maintain a spherical particle in motion along a prescribed elliptic path. The path is given as $\mathbf{Y}(t) = \mathrm{Re}\left[\hat{\mathbf{Y}}\exp(i\sigma t)\right]$, with a similar expression for the velocity $\mathbf{V}(t)$. The required external force is then $\mathbf{F}^{ext}(t) = \mathrm{Re}\left[\hat{\mathbf{F}}^{ext}\exp(i\sigma t)\right]$, where

$$\hat{\mathbf{F}}^{ext} = \left(m_P + \tfrac{1}{2}m_F\right)i\sigma\hat{\mathbf{V}} + 6\pi a\mu\left(1 + \kappa e^{i\pi/4}\right)\hat{\mathbf{V}} \tag{9}$$

and $\kappa^2 = \sigma a^2/\nu$. The value of κ indicates the relative importance of the Basset history term and the ratio of this force to that of added mass is equal to $9/\kappa$. Indeed at lower frequencies σ, this history term is more important than added mass as a correction to the particle motion. Another point to note is the direction of the forces relative to the particle velocity. While the Stokes drag force simply acts in opposition to the particle velocity relative to any ambient flow, both the added mass and history term act at angles to this. With a circular path, the effects of particle inertia or added mass act orthogonally to the particle's path while the time lag in the viscous force, through the history term, acts at 45° to the path.

Various simplifications to (6) are often used for practical reasons, such as neglecting the unsteady viscous response of the Basset term. The latter can be difficult to compute in a numerical solution and requires the retention of the particle acceleration or $d\mathbf{Q}/dt$ over a long period of time. Efficient schemes for numerically computing the Basset term have been proposed (van Hinsberg et al. 2011) and more numerical methods have been developed recently to solve fractional-order, differential equations in general. As we consider the effects of finite Reynolds number on the particle motion there is good reason to truncate the effect of the Basset term.

2.3 *Effects of Finite Particles Reynolds Number*

These results for very low Reynolds numbers are informative and useful but they cover a limited range. More usually the Reynolds numbers associated with the particle motion, Re_P and Re_Γ, are not zero. This introduces other physical effects even if the Reynolds numbers are still small or $O(1)$.

Consider first the simple steady motion of a sphere in the absence of any other flow. There is an increase in the viscous drag force, relative to the Stokes value, given by the Oseen correction so that the external force needed to sustain a particle velocity $\mathbf{V}$ is

$$\mathbf{F}^{ext} = 6\pi\mu a\mathbf{V}\left(1 + \tfrac{3}{16}Re_P\right) \tag{10}$$

The usual assumption that (5) reduces to a viscous Stokes flow for $\mathrm{Re}_P \ll 1$ does not hold at large distances from the sphere, see Batchelor (1967). Eventually in the far field, where $r \sim a\mathrm{Re}_P^{-1}$, the effect of flow advection must be considered and relative to the sphere there is an axisymmetric wake structure. The specific result (10) of Oseen (1910) is obtained by taking the sphere to be fixed and assuming a uniform far field flow $\mathbf{U}$ past the sphere while the sphere itself is viewed simply as a point force at the origin. This gives the Oseenlet response to a point force, as opposed to the more usual Stokeslet for Stokes flow. The final result (10) comes from matching the outer Oseenlet flow to the near-field Stokes flow past the sphere. This is summarized by Guazzelli and Morris (2012, Chap. 8), where the mass flow balance within the wake is also described.

These corrections for the drag force in steady motion only apply for small Re_P and beyond this one must rely on empirical correlations for the drag force or results from direct numerical simulations of the flow. Figure 3 shows the streamlines for steady, axisymmetric flow past an isolated sphere at $\mathrm{Re} = 2aU/\nu = 10$ and 40 with intervals

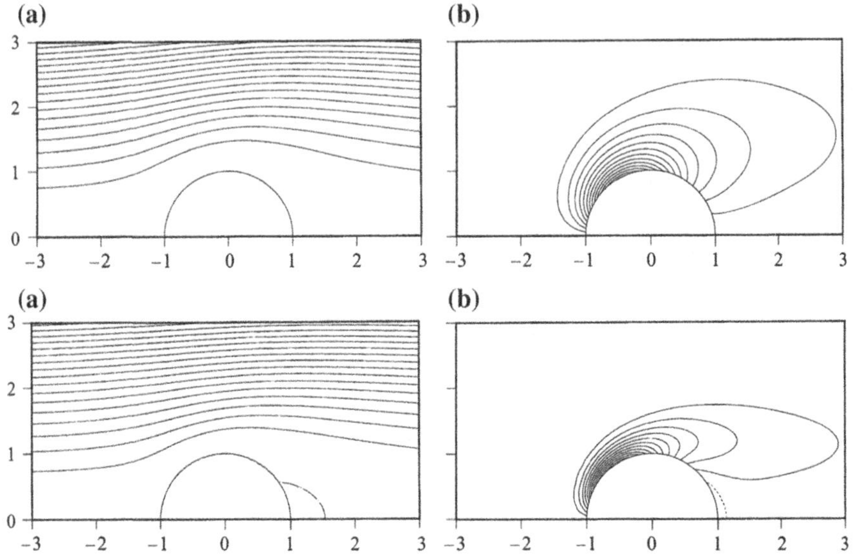

Fig. 3 Steady flow past a fixed sphere at Re $= 2aU/\nu = 10$ (*top*) and 40 (*bottom*), showing the streamlines **a** and vorticity contours **b**, from Chang and Maxey (1994), reproduced with permission

of $\Delta\psi = 0.25$, as scaled by U and a. Similarly the vorticity contours are shown at intervals of 0.25. The results are from the computations by Chang and Maxey (1994). Immediately evident is the asymmetry of the flow from left to right as compared to a simple Stokes flow. The vorticity contours are swept downstream even near to the sphere and not just in the far field as in the Oseen wake. Close to Re $= 20$, an attached separated flow develops. At Re $= 40$ there is a small region of negative vorticity close to the body surface, shown by the dashed line contour. Surface vorticity is important as the viscous stresses on a rigid sphere can be determined directly from this. As the Reynolds number increases further the flow remains axisymmetric for Re < 210, but becomes inherently three-dimensional and eventually unsteady at higher Reynolds numbers (Tomboulides and Orszag 2000; Jenny et al. 2004).

If the ambient flow past the sphere remains unidirectional but is now unsteady, the local flow is still axisymmetric for low to moderate Reynolds numbers but one must reconsider the nature of the added mass effect and other fluid forces. Numerical computations of the flow can provide some basic answers. Figure 4 shows the streamlines and vorticity contours for oscillatory flow past a sphere at different phases of the oscillation cycle. The peak Reynolds number is 16.7, less than the value of 20 where one may see separation in steady flow, yet as the flow decelerates instantaneous separation is seen in frame (d). This is a common feature of oscillating flows where local flow reversal will occur at different locations on the sphere surface as the oscillation phase varies. Only at zero Reynolds number, for unsteady Stokes flow, is there a simultaneous flow reversal on the sphere. An oscillating flow is characterized by both

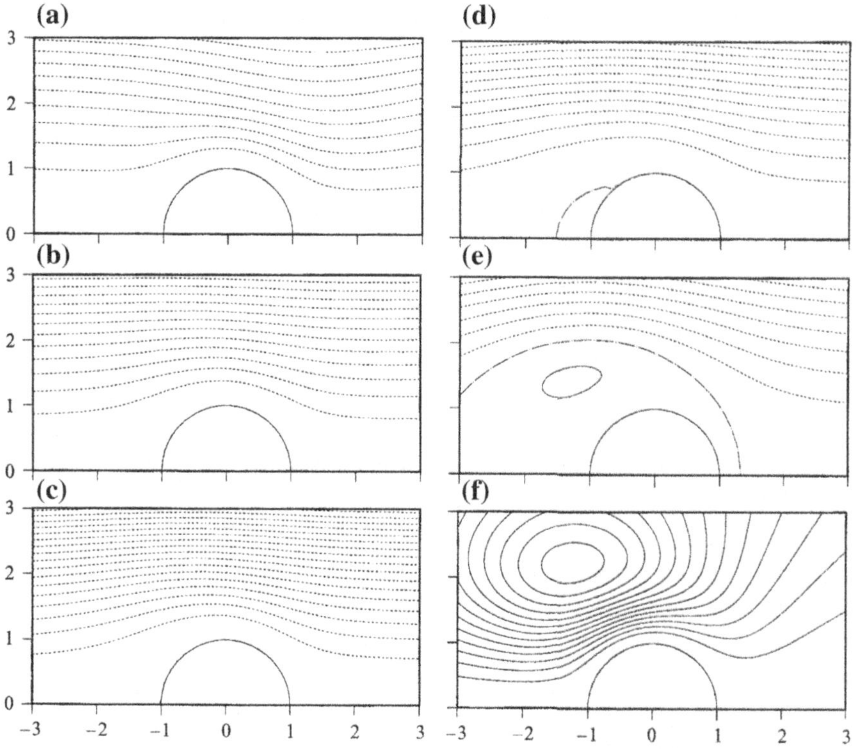

Fig. 4 Streamlines for an oscillating flow $U(t) = -X\sigma \sin(\sigma t)$ past a sphere, shown for $X = 1.6a$ and Re $= 16.7$, over half a cycle at **a** $\sigma t = \pi/16$, $\Delta\psi = 0.1$, **b** $\sigma t = \pi/4$, $\Delta\psi = 0.25$, **c** $\sigma t = \pi/2$, $\Delta\psi = 0.25$, **d** $\sigma t = 3\pi/4$, $\Delta\psi = 0.25$, **e** $\sigma t = 15\pi/16$, $\Delta\psi = 0.1$, **f** $\sigma t = \pi$, $\Delta\psi = 0.025$. Values of the streamfunction ψ are given in intervals of $\Delta\psi$; positive ψ (*solid lines*), negative ψ (*broken line*), and $\psi = 0$ (*dashed dotted line*). From Chang and Maxey (1994), reproduced with permission

the peak Reynolds number and a second parameter such as the relative oscillation amplitude X/a, or the parameter κ.

The specific question about the general form of the added mass for a sphere was addressed by Auton et al. (1988) and it was argued that this force is

$$\mathbf{F}^{AM} = \frac{m_F}{2}\left(\frac{d\mathbf{V}}{dt} - \frac{D\mathbf{u}}{Dt}\right) \tag{11}$$

In the context of low Reynolds numbers, this form is approximately the same as in (6), without the correction for the finite particle size and nonuniform ambient flow. The added mass is closely associated with the pressure response as already noted. Several studies have been made using numerical simulations to verify this result even in contexts where there may be an attached region of separated flow. Both sharp accelerations and decelerations of the flow in time (Rivero et al. 1991; Chang and Maxey 1995) and the force on a sphere in a linear straining flow (Magnaudet et al.

1995; Bagchi and Balachandar 2003) verify that (11) is correct for unsteady finite Reynolds number flows.

The nature of the unsteady fluid forces on a sphere that is in an accelerating, uniform flow has been investigated both in numerical simulations and theoretically, see for example Sano (1981), Mei and Adrian (1992), Lovalenti and Brady (1993), Lovalenti and Brady (1995), Lawrence and Mei (1995) and the references already mentioned. Theoretical work has focused on the low Reynolds number range for which an Oseen approximation is appropriate. A key observation is that the viscous Basset term, and specifically the kernel within the integral, decays more quickly than the usual $t^{-1/2}$ in the long term. The Basset form is appropriate for times $t < O(a^2/\nu)$, but at later times the advection of vorticity in the Oseen wake becomes more important than simple viscous diffusion of vorticity alone. The exact response depends on whether the flow is accelerating from rest or if there is a large or small step change in the velocity. These estimates for final decay of the unsteady viscous force ΔF may be summarized as follows, with t scaled by a^2/ν, see Lovalenti and Brady (1993, Appendix D).

- Start from rest, $\Delta F \sim t^{-2}$
- An established flow is brought to rest, $\Delta F \sim t^{-1}$
- Small change in the flow, $\Delta F \sim t^{-5/2}\exp(-\alpha t)$
- Large change in the flow, $\Delta F \sim t^{-2}\exp(-\beta t)$

The constants α, β depend on whether the flow accelerates or decelerates.

These results and those from numerical simulations have lead to several proposed generalizations of the Basset history integral for accelerated motion in a uniform flow as,

$$\mathbf{F}_H = 6\pi\mu a \int_0^t K(t-\tau)\frac{d(\mathbf{V}-\mathbf{u})}{d\tau}d\tau. \tag{12}$$

Several of the possible forms of the kernel $K(t-\tau)$ are given by Mei and Adrian (1994), Loth and Dorgan (2009), Ling et al. (2013) and references therein. In general numerical simulations show too that the viscous drag force comes to an equilibrium value more quickly for a range of non-zero Reynolds numbers than would be indicated by the Basset result. At this late stage the adjustment in the particle velocity may be 10% or less.

2.4 *Lift Forces*

So far we have considered the effects of non-zero particle Reynolds number in a spatially uniform flow. The presence of a velocity gradient in the ambient flow introduces additional effects at finite Reynolds number, unlike Stokes flow where there was no change in the fluid force on a sphere. At low Reynolds numbers, a shear flow will distort the otherwise axisymmetric, outer Oseen wake and both the vorticity and pressure distributions near the sphere are altered. The resulting fluid

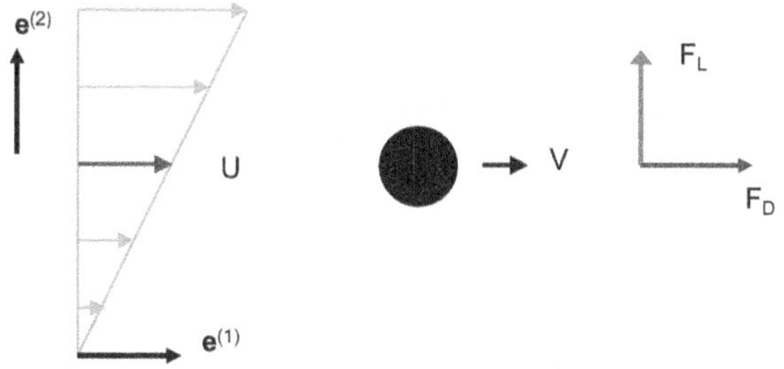

Fig. 5 Sketch of a particle moving in a simple shear flow, for $W = U - V > 0$, and the resulting lift and drag forces

force is no longer aligned with the ambient flow relative to the velocity of the particle, $\mathbf{W} = (\mathbf{u}(\mathbf{Y}(t), t) - \mathbf{V}(t))$. There is instead a drag force component $\mathbf{F}_D$ aligned with the relative velocity $\mathbf{W}$ and a so-called lift force $\mathbf{F}_L$ orthogonal to this. The result is that a lift force may be generated as illustrated in Fig. 5.

Saffman (1965), see also Saffman (1968), derived a theoretical estimate for $\mathbf{F}_L$ by modifying the usual Oseen flow analysis for low Reynolds number conditions to include a uniform shear flow in the outer ambient flow. In steady motion, the disturbance flow $\mathbf{w}^{(1)}$ will now satisfy the conditions of incompressible flow, $\nabla \cdot \mathbf{w}^{(1)} = 0$ and

$$(-\mathbf{W} + (\Gamma \cdot \mathbf{z})) \cdot \nabla \mathbf{w}^{(1)} + \Gamma \cdot \mathbf{w}^{(1)} + \mathbf{w}^{(1)} \cdot \nabla \mathbf{w}^{(1)} = -\nabla(p'/\rho) + \nu \nabla^2 \mathbf{w}^{(1)} \quad (13)$$

The boundary conditions are that $\mathbf{w}^{(1)} = 0$ as $|\mathbf{z}| \to \infty$ and on the surface of the sphere at $|\mathbf{z}| = a$ the flow is $\mathbf{w}^{(1)} = -\mathbf{W} + \boldsymbol{\Omega} \times \mathbf{z}$. The ambient flow $\mathbf{u}$ is a simple uniform shear flow,

$$\mathbf{u} = \mathbf{u}(\mathbf{Y}(t), t) + \Gamma \cdot \mathbf{z} = (U + \beta z_2)\mathbf{e}^{(1)} \quad (14)$$

and is itself a solution of the Navier–Stokes equations. The process of matched asymptotic expansions for the inner region, where $|z| = O(a)$, gives an initial estimate for the fluid force on the particle as a simple Stokes drag $\mathbf{F}_D = 6\pi\mu a\mathbf{W}$. In the outer region, Saffman assumed that the shear rate β was dominant with $\mathrm{Re}_P \ll \mathrm{Re}_\Gamma \ll 1$ so that in the region where $|z| = O(a)\mathrm{Re}_\Gamma^{-1/2}$ the first correction to the flow comes from solving the Oseen problem

$$\beta z_2 \frac{\partial}{\partial z_1} \mathbf{w}^{(1)} + \beta w_2^{(1)} \mathbf{e}^{(1)} = -\nabla(p'/\rho) + \nu \nabla^2 \mathbf{w}^{(1)} - 6\pi\mu a \mathbf{W} \delta(\mathbf{z}) \quad (15)$$

At this scale, the sphere appears as a localized point force acting on the fluid which in turn generates a correction to the fluid velocity $\mathbf{w}^{(1)}$ at $\mathbf{z} = 0$. One may view the particle as acquiring a drift velocity $\mathbf{V}_L = 0.343\mathrm{Re}_\Gamma^{1/2} W \mathbf{e}^{(2)}$ or a lift force

$$\mathbf{F}_L = 6.46 a\mu \mathrm{Re}_\Gamma^{1/2} W \mathbf{e}^{(2)} \quad (16)$$

A good summary of Saffman's results is provided by Stone (2000).

There are many variations to this problem. For example Rubinow and Keller (1961) used a similar Oseen flow and a matched asymptotic procedure to evaluate the lift force on a sphere moving with velocity **V** and rotating with angular velocity $\boldsymbol{\Omega}$ in otherwise still fluid. The lift force is

$$\mathbf{F}_L = \rho\pi a^3 \boldsymbol{\Omega} \times \mathbf{V}(1 + O(\mathrm{Re}_P)), \tag{17}$$

with the drag force as given by (10). At low Reynolds number this lift force is smaller in magnitude than the Saffman lift force, although both act in the same direction, for a sphere freely rotating with the local vorticity.

Numerical simulations have supported these estimates and evaluated the forces over a range of Reynolds numbers Re_P and Re_Γ. Bagchi and Balachandar (2002) review much of the prior work and give numerical results for a sphere in a uniform shear flow. They discuss the effect of whether the sphere is fixed or freely rotates in the shear flow. In the end the data provides a set of correlations for $\mathbf{F}_L$ as a function of Re_P and Re_Γ. McLaughlin (1991) extended Saffman's analysis to show that as the ratio of $\mathrm{Re}_\Gamma^{1/2}$ to Re_P becomes smaller the lift force is reduced and may actually change sign. The lift force is also modified by the presence of a nearby planar wall, as may occur in a Couette flow or Poiseuille flow (McLaughlin 1993; Asmolov 1999).

This discussion of effects arising at finite particle Reynolds number shows that there are many possible responses of the particle motion that one should consider. Trying to capture all of these in a single equation of motion such as (6) is a daunting task. If we do use such an equation of motion we should be aware of the limitations. Nevertheless this can be a productive exercise to gain a first idea as we explore possible particle motions.

3 Isolated Particles in Simple Flows

Studying particle motion in simple, nonuniform shear flows is a valuable step in understanding the effects of turbulence or other complex unsteady flows on particle transport. Even if the particles are well separated and their mass loading is very low so that the bulk flow is not altered, the distribution of the particles can be substantially affected by the flow. We consider as a basic, two-dimensional incompressible flow the periodic array of vortices given by the streamfunction

$$\psi(x_1, x_2) = U_0 L \sin\left(\frac{x_1}{L}\right) \sin\left(\frac{x_2}{L}\right) \tag{18}$$

We choose U_0 and L as the reference scales so that the velocity components are

$$\begin{aligned} u_1 &= \sin(x_1)\cos(x_2) \\ u_2 &= -\cos(x_1)\sin(x_2) \end{aligned} \tag{19}$$

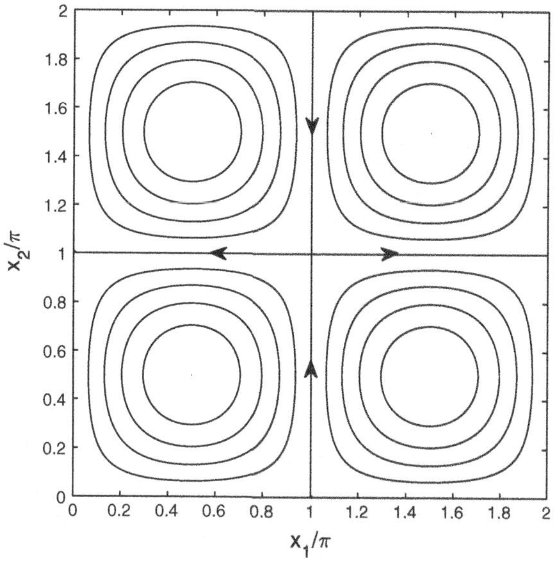

Fig. 6 Streamlines showing the periodic array of vortices of alternating sign given by (18). The *arrows* show the direction of the flow

This flow is illustrated in Fig. 6, where the streamlines are shown for the range $0 \leq x_1/\pi, x_2/\pi \leq 2$.

This flow pattern arises in many contexts such as the onset of Rayleigh–Benard thermal convection with free-slip boundary conditions (Drazin and Reid 2004) and was used by Stommel (1949) to consider the suspension of marine organisms by convective motions near the sea surface. It is also sometimes referred to as the 2-D Taylor–Green vortex, after the early work on a corresponding 3-D flow used to study formation of a vorticity cascade in turbulence (Taylor and Green 1937). The flow is characterized by strong vorticity at the center of each cell and stagnation points with pure straining at the corners. It is an exact solution of the Euler equations for inviscid flow and in a viscous fluid U_0 decays slowly, but the flow (18) maintains the same form. The flow can be generated in the lab through Lorentz body forces in a conducting fluid layer above an array of permanent magnets of alternating polarity (Kelley and Ouellette 2011; Bergougnoux et al. 2014). For theoretical work, it is relatively simple to compute the motion of different particles using standard numerical methods.

The first example is for the motion of solid particles in a gas flow, according to (4) where there is negligible particle inertia, $St = U_0\tau_P/L = 0$. While Lagrangian fluid tracers ($W_S = 0$) will simply move along the closed streamlines of Fig. 6, a settling particle ($-W_2 = W = W_S/U_0 > 0$) may either be permanently suspended in a closed trajectory or fall vertically along an open path. The closed paths are driven by an upflow over one segment that is greater than W and then the particle falls and is swept by the flow in a loop over the rest of the path. The fraction of suspended particles varies from 100% at $W = 0$ to 0% if $W \geq 1$. This is illustrated in Fig. 7 and was noted by Stommel. A key feature is that if particles are uniformly dispersed in

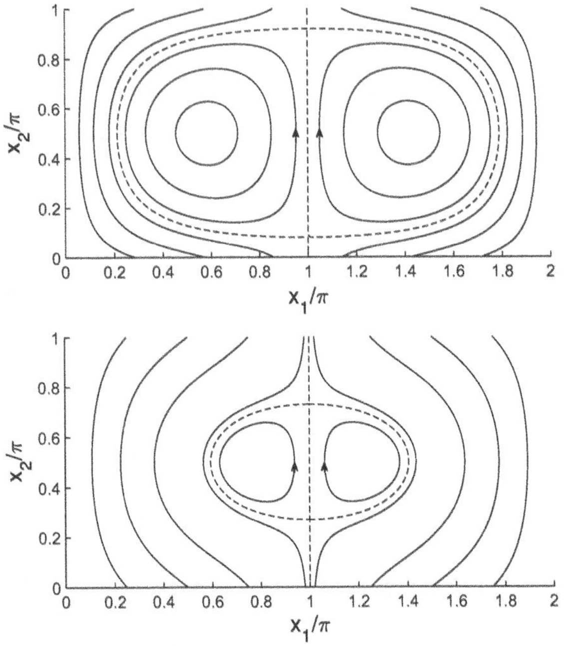

Fig. 7 Trajectories of non-inertial particles, $St = 0$, in the vortex flow as given by ψ_P, for (*upper*) $W = 0.25$, where 60% particles are suspended; (*lower*) $W = 0.75$, 18% are suspended. Bounding pathline $\psi_P = W\pi$ shown as *dashed curve*

the flow at the outset, they remain so. One can also define a particle streamfunction for this motion, $\psi_P = \psi + W_1 x_2 - W_2 x_1$ giving $\mathbf{V}(t)$ in terms of a particle velocity field $\mathbf{v}(\mathbf{x}, t)$ evaluated at $\mathbf{x} = \mathbf{Y}$.

This situation changes if we introduce particle inertia, solving (3) with $St > 0$. Both initial positions and velocities must be specified, and here we set the initial velocity to be zero. Figure 8 shows the positions at $t = 100L/U_0$ for particles that are initially arranged on a uniform grid in each cell, with $W = 0.25$. Remarkably, all the particles are now settling and follow open paths. Secondly, the different trajectories eventually all merge into isolated preferred paths. For $St = 0.2$, the trajectories merge into a single path in each cell and at all points are swept by the flow in a downward direction. The overall settling rate is increased. For $St = 1$ there is more particle inertia. These particles are also swept along by the flow and all eventually settle. There is again a merging into isolated paths, but these now have a more complex structure. This becomes clearer if we examine the later stages of the paths for some individual particles, as shown in Fig. 9. The paths are periodic in the vertical but the period is now 4π instead of 2π. The particles cross vertical cell boundaries and then cross back again. The scatter plot of Fig. 8 shows an overlay of four basic trajectories that pass through the cells shown, corresponding to different phases.

These and other results are described by Maxey and Corrsin (1986), Rubin et al. (1995). It should be noted that an alignment of the cells with the vertical can create special patterns where particles collect on the vertical cell boundaries if $W > 1$ and it is worth looking at the motion with both W_1 and W_2 non-zero.

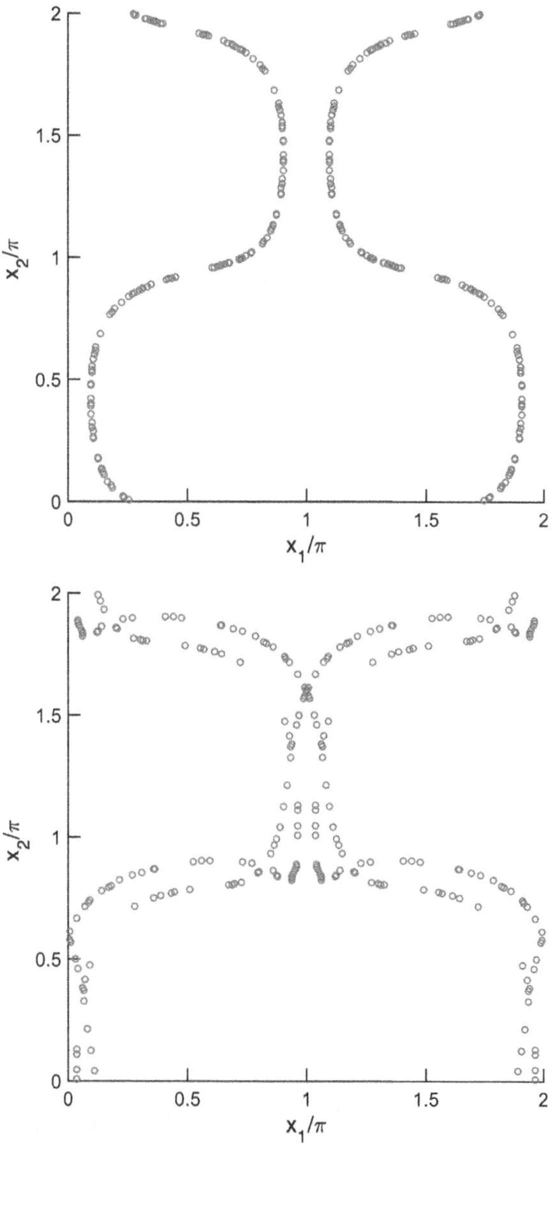

Fig. 8 Scatter plot at $t = 100$ for the positions of inertial particles evenly dispersed at $t = 0$ on a 8×8 lattice in each cell, $W = 0.25$: $St = 0.2$ (*upper*); $St = 1$ (*lower*)

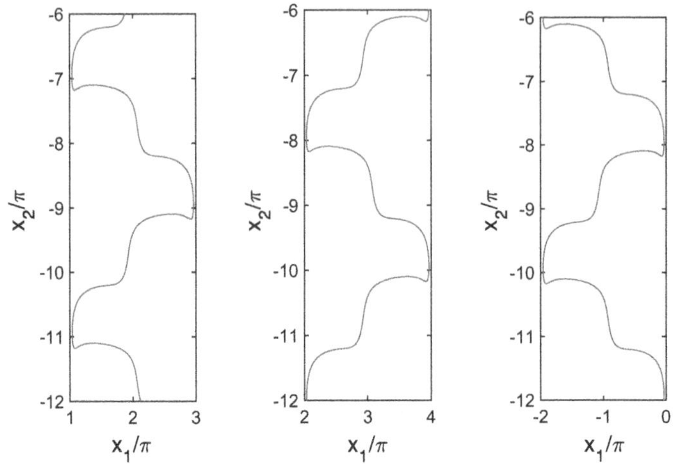

Fig. 9 Selected particle trajectories for $W = 0.25$ and $St = 1$. Particles start from rest at $Y_2 = 1.8\pi$, within the interval $0 < Y_1/\pi < 2$

A specific feature of the $St = 1$ results is that the trajectories can cross, so that a particle observed at a specific location may have different velocities depending on its past motion. This can be understood further by considering a dense inertial particle in a simple, stagnation point flow where $u_1 = \alpha x_1$, $u_2 = -\alpha x_2$. The solution to (3) for the position $\mathbf{Y}(t)$ can be written as

$$Y_1(t) = +B_1 \exp \lambda_1 t + C_2 \exp \lambda_2 t$$
$$Y_2(t) = -\frac{1}{\alpha} W + B_2 \exp \lambda_1 t + C_2 \exp \lambda_2 t \tag{20}$$

where λ_1 and λ_2 are the roots

$$\lambda = \{-1 \pm \sqrt{1 - 4\alpha\tau_P}\}/(2\tau_P) \tag{21}$$

At $\mathbf{Y} = (0, -W/\alpha)$ there is an equilibrium point, where the particle is suspended by the upflow. If $St = \alpha\tau_P < 0.25$, there is no overshoot of the line $x_2 = -W/\alpha$ for a particle that is initially above this line. However for $St > 0.25$, the roots for λ are a complex conjugate pair and overshoot can occur in either direction for particles starting above or below this line. Thus we can expect that for low values of St there will be no crossing of these asymptotic trajectories. Even with the crossing of trajectories at $St \sim 1$, we still find a clustering or merging of the particle trajectories in the periodic flow (18).

This analysis for low St can be taken further and an asymptotic approximation for (3) made as

$$\mathbf{V} = \mathbf{v}(\mathbf{Y}, t) = \mathbf{u}(\mathbf{Y}, t) + \mathbf{W}_S - St\left(\frac{\partial \mathbf{u}}{\partial t} + (\mathbf{u} + \mathbf{W}) \cdot \nabla \mathbf{u}\right), \tag{22}$$

where the derivatives are evaluated at $\mathbf{Y}(t)$. This result corresponds to a specific choice of the initial velocity of the particle but this has limited effect on the long term trajectory. The point is that for low St the particle velocity can be written as a specific function of the particle position. A consequence is that for a non-settling ($W = 0$), inertial particle the trajectory will converge to the streamline $\psi = 0$ where the particle moves along a cell boundary (Maxey and Corrsin 1986). Another feature is that this flow field $\mathbf{v}(\mathbf{x}, t)$ is compressible. Inertia makes it harder for particles to turn corners as it follows the flow, causing the particles to drift out of a vortex core and towards a region of higher strain rate. This has become termed preferential concentration of inertial particles and much has been written on the subject. The effects of particle inertia and these dynamics on rates of particle settling in turbulence were considered by Maxey (1987a) and by Wang and Maxey (1993) and a more recent overview is given by Monchaux et al. (2012). Balachandar and Eaton (2010) give additional information on the effects of preferential concentration of particles in turbulent flows.

As the effects of inertia become stronger, when $St \geq 1$, other features have been noted for particles in turbulent or random flow fields. One is the formation of caustics among the trajectories of different particles that is linked to the crossing of their paths, see Wilkinson et al. (2007) and references therein. With large inertia, the particles have a more limited response to the local flow and can travel along nearly straight paths for some distance before being caught in some region long enough for the flow to reorient the motion (Ijzermans et al. 2010). This can be seen with the present periodic flow (18) if one looks at the motion with tilted cells, $W_1 \neq 0$. For example if $W = 0.25$, $St = 10$ and the cells are tilted slightly at an angle of 15° to the vertical, the trajectories are reminiscent of a pinball machine.

This approach of looking at particle motion in a simple, nonuniform flow may be applied to solid particles or gas bubbles in liquid flows or to nonspherical particles such as ellipsoids. A first idea of the response in a solid in liquid flow may be obtained by computing the trajectories with a simplified version of (6),

$$\left(m_P + \tfrac{1}{2}m_F\right)\frac{d\mathbf{V}}{dt} = (m_P - m_F)\,\mathbf{g} + 6\pi a\mu\,(\mathbf{u} - \mathbf{V}) + \tfrac{3}{2}m_F\frac{\partial \mathbf{u}}{\partial t} + m_F\left(\mathbf{u} + \tfrac{1}{2}\mathbf{V}\right)\cdot\nabla\mathbf{u}. \tag{23}$$

This retains the effect of added-mass and buoyancy but neglects history terms and other effects. The scaled version of this equation is

$$\frac{d\mathbf{V}}{dt} = \frac{1}{St}(\mathbf{u} - \mathbf{V} + \mathbf{W}) + \tfrac{3}{2}R\frac{\partial \mathbf{u}}{\partial t} + R(\mathbf{u} + \tfrac{1}{2}\mathbf{V})\cdot\nabla\mathbf{u} \tag{24}$$

The parameter R is based on the ratio of the particle mass to the mass of displaced fluid as $R = m_f/(m_P + 0.5m_F)$. Thus $R = 0$ for a solid particle in a gas flow, $R = 2/3$ corresponds to a neutrally buoyant particle. $R = 2$ would represent a contaminated gas bubble, where $m_P = 0$ but the effect of surfactants immobilizes the liquid on the bubble surface giving a no-slip response. Similarly the definition of Stokes number

Fig. 10 Sample trajectories of rising bubbles, $St = 0.2$, for $W = 0.75$ (*upper*); and corresponding scatter plot (*lower*)

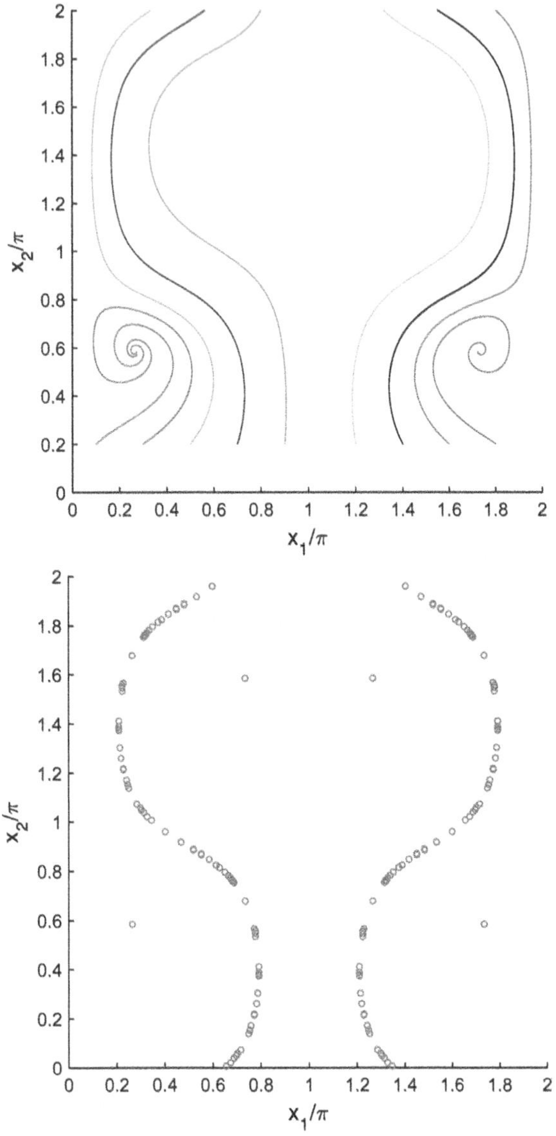

is based on $\tau_P = (m_P + 0.5m_F)/6\pi a\mu$ and $\mathbf{W}$ is the scaled terminal fall or rise velocity in still fluid.

Some representative results for the motion of small gas bubbles, $R = 2$, are shown in Fig. 10 for $W = 0.75$. There is a small degree of inertia, $St = 0.2$, as appropriate for small bubbles or solids in liquids. Some of the bubbles become trapped at fixed points, which unlike the gas–solid case, are now stable. These are equilibrium points where the local downflow balances the buoyant rise of the bubble. A substantial fraction of the bubbles may be trapped in this way. Other bubbles follow open paths rising through successive cells. Trapping of the bubbles is more prevalent at smaller values of W or for larger St. If we examine the corresponding scatter plot at $t = 100$, we see that the freely rising bubbles merge into isolated paths as found for the inertial gas–solid particles but now moving in the upflow region of each cell. Referring back to (22), the corresponding result is that buoyant particles tend to collect in regions of stronger vorticity and lower strain rate (Maxey 1987b). The recent work by Bergougnoux et al. (2014) explores this system in much more detail and compares experimental measurements and simulations, taking fuller account of the fluctuating fluid forces on the particles. Their focus is on the very interesting issues of how particles respond close to neutral buoyancy, $R = 2/3$.

If one considers the motion of small spheroidal particles, there are further degrees of freedom and interesting dynamics even without the effects of inertia. Jeffery (1922) showed that a spheroid in viscous flow will tend to rotate with local vorticity, much as a sphere does, but also will tend to align so that the long axis is in the direction of largest principal strain. A spheroid placed in a uniform shear flow exhibits the Jeffery orbits, whereby it turns at different rates depending on its orientation in the flow. Further, the resistance to motion along the long axis of a prolate spheroid may be half that transverse to the axis. This means that a spheroid may not settle vertically under gravity, depending on its orientation. If this is coupled to motion in a nonuniform flow (18) some unusual features emerge. One is that the particle motion may be chaotic as it tumbles and settles under gravity. There are regions of regular motion in the flow, quasi-periodic trapping or suspension, and chaos (Mallier and Maxey 1991; Shin and Maxey 1997).

4 Force Coupling Method

Many of the current research challenges involve suspensions of particles that are not isolated and interact with each other or with flow boundaries. The finite size of the particle is an issue and commonly the particle and fluid are of comparable density. As we noted at the end of Sect. 2, it is difficult to develop an equation of motion that encompasses all these conditions even for low Reynolds number systems. This necessitates some form of numerical simulation of the problem posed by (5) for the disturbance flow generated by a system of particles, the fluid forces that these generate $\mathbf{F}^{(F)}$ and a corresponding momentum equation for each particle,

$$m_P \frac{d\mathbf{V}}{dt} = \mathbf{F}^{(F)} + \mathbf{F}^{(Ext)}. \tag{25}$$

A similar equation is specified for the angular velocity $\boldsymbol{\Omega}$ in terms of the fluid and external torques $\mathbf{T}^{(F)}$, $\mathbf{T}^{(Ext)}$ on the particles and the moment of inertia I_P of the particle.

There are several ways that this has been approached. The arbitrary Lagrangian Eulerian (ALE) scheme of Hu et al. (2001) employs a body-fitted numerical mesh around each particle and solves for the incompressible flow with a finite element scheme. The mesh moves with the particles, but must be rebuilt and the data interpolated to the new mesh at intervals as it becomes increasingly distorted. This can be very accurate but is challenging to scale up to large systems. More common is some form of fictitious domain method that is based on a fixed computational mesh for the flow and fictitious forces are applied to the fluid to mimic the presence of a rigid body. Examples of this are the distributed Lagrangian multiplier (DLM) scheme Glowinski et al. (1999) and the immersed boundary method (IBM) (Uhlmann 2005; Breugem 2012). Here we outline an effective, yet simplified scheme based on a distribution of low-order finite force multipoles acting on the fluid.

The force coupling method (FCM) is a fictitious domain scheme whereby the presence of a particle in a flow $\mathbf{u}(\mathbf{x}, t)$ is represented by a finite force monopole and force dipole,

$$\rho_F \left(\frac{\partial u_i}{\partial t} + u_j \frac{\partial u_i}{\partial x_j} \right) = -\frac{\partial p}{\partial x_i} + \mu \nabla^2 u_i +$$

$$\sum_n F_i^n \Delta_M(\mathbf{x} - \mathbf{Y}^n) + \sum_n G_{ij}^n \frac{\partial \Delta_D}{\partial x_j}(\mathbf{x} - \mathbf{Y}^n). \quad (26)$$

The functions Δ_M and Δ_D for the monopole and dipole are given by

$$\Delta(\mathbf{x}) = (2\pi\sigma^2)^{-3/2} \exp(-\mathbf{x}^2/2\sigma^2), \quad (27)$$

where for a sphere $a/\sigma_M = \sqrt{\pi}$ and $a/\sigma_D = (6\sqrt{\pi})^{1/3}$. This is posed as a mobility problem. The strength of the force monopole F_i is specified as

$$\mathbf{F} = \mathbf{F}^{(Ext)} + (m_F - m_P)\frac{d\mathbf{V}}{dt} + \mathbf{F}^{(Con)} \quad (28)$$

This replaces the momentum of the rigid body by the corresponding momentum of a rigid fluid sphere moving with same velocity $\mathbf{V}$ resulting in a force $\mathbf{F}$ of the sphere on the fluid. We introduce a short-range contact force $\mathbf{F}^{(Con)}$ that acts between particles to prevent unphysical overlaps of the rigid bodies. The particle velocity $\mathbf{V}$ is determined from the resulting flow,

$$\frac{d\mathbf{Y}}{dt} = \mathbf{V}(t) = \int \mathbf{u}(\mathbf{x}, t) \Delta_M(\mathbf{x} - \mathbf{Y}(t)) d^3\mathbf{x} \quad (29)$$

The basic details of the method are given in a series of papers (Maxey and Patel 2001; Lomholt and Maxey 2003; Liu et al. 2002).

The aim is to match the low-order integral moments of the flow to those that would apply to the true motion of a rigid body in the fluid. The mobility format breaks from the more usual resistance model and ensures a tighter coupling of particle and fluid momentum. The angular velocity $\boldsymbol{\Omega}$ is found in a similar manner from the fluid vorticity as

$$\Omega_i(t) = \tfrac{1}{2} \int \omega_i(\mathbf{x}, t)\Delta_D(\mathbf{x} - \mathbf{Y}(t))d^3\mathbf{x} \tag{30}$$

The force dipole coefficient G_{ij} has a symmetric part G_{ij}^S, commonly referred as a stresslet, and an anti-symmetric part G_{ij}^A that may be linked to a torque acting on the fluid, $G_{ij}^A = (1/2)\epsilon_{ijk}T_k$. The value of the torque can be set as

$$\mathbf{T} = \mathbf{T}^{(Ext)} + (I_F - I_P)\frac{d\boldsymbol{\Omega}}{dt}, \tag{31}$$

with an appropriate value for the fluid moment of inertia I_F.

The symmetric stresslet term G_{ij}^S is set indirectly so as to ensure that the volume-averaged rate of strain s_{ij} for each particle is zero, thus

$$\int \frac{1}{2}\left(\frac{\partial u_i}{\partial x_j} + \frac{\partial u_j}{\partial x_i}\right) \Delta_D(\mathbf{x} - \mathbf{Y}(t))d^3\mathbf{x} = 0 \tag{32}$$

This condition is certainly consistent with the concept of the fluid inside particle volume responding as a rigid body. It also ensures that the stresslet does no net work. One can evaluate the budget for the total kinetic energy in the flow domain based on (26). For particles suspended in a flow domain D, the rate of change of the total kinetic energy is

$$\begin{aligned} \frac{d}{dt}\int_D \tfrac{1}{2}\rho\mathbf{u}^2 d^3\mathbf{x} &= \oint_{\partial D} u_i\tau_{ij}n_j dS - \int_D 2\mu s_{ij}s_{ij}d^3\mathbf{x} \\ &+ \sum_n F_i^{(n)} \int_D u_i\Delta_M(\mathbf{x} - \mathbf{Y}^{(n)})d^3\mathbf{x} + \sum_n T_k^{(n)}\Omega_k^{(n)} \end{aligned} \tag{33}$$

The second to last term of (33) can be written as $\sum_n F_i^{(n)}V_i^{(n)}$. The last two terms relate to the work done by the external forces and torques and the change in kinetic energy of the rigid particles and of the fictitious fluid inside the particles. The other terms on the right hand side are the rate of work by forces on the domain boundary and the rate of viscous dissipation. The key point is that this is evaluated over the whole domain, not just the region outside the particles. The fictitious fluid motion inside the particle is dynamically relevant and as will be shown the FCM represents a local, spatial smoothing of the dynamics.

4.1 Stokes Flows

The FCM representation for the Stokes flow of a sphere moving steadily in otherwise still fluid under the action of an external force **F** can be found analytically. The solution of (26) with a simple monopole force consists of three parts,

$$u_i = \left(S_{ij}^{(1)} + S_{ij}^{(2)} + S_{ij}^{(3)} \right) F_j. \tag{34}$$

These terms are,

$$S_{ij}^{(1)} = \frac{1}{8\pi\mu r}\left(\delta_{ij} + \frac{x_i x_j}{r^2}\right) \operatorname{erf}(r/\sigma\sqrt{2}), \tag{35}$$

$$S_{ij}^{(2)} = \frac{1}{8\pi\mu r^3}\left(\delta_{ij} - 3\frac{x_i x_j}{r^2}\right) \sigma^2 \operatorname{erf}(r/\sigma\sqrt{2}), \tag{36}$$

$$S_{ij}^{(3)} = -\frac{\sigma^2}{2\mu}\left(\delta_{ij} - 3\frac{x_i x_j}{r^2}\right) \frac{\sigma^2}{r^2} \Delta_M(\mathbf{x}). \tag{37}$$

The first term $S_{ij}^{(1)}$ matches a classic Stokeslet flow due to a point-force **F** once r/σ is large enough that $\operatorname{erf}(r/\sigma\sqrt{2}) \sim 1$. The second term $S_{ij}^{(2)}$ similarly corresponds to a potential dipole flow in the far field while the third term is a localized flow within the sphere volume that ensures that the total flow is incompressible.

Without making an assumption about the value of σ, we can evaluate (29) as $V_i = Q_0 F_i/\mu\sigma$ and the value of Q_0 is $Q_0^{-1} = 6\pi^{3/2}$. This matches the Stokes drag resistance if we set $a/\sigma_M = \sqrt{\pi}$. The exact result for Stokes flow past a rigid sphere of radius a moving with velocity **V** is

$$u_i = \frac{3}{4}\frac{aV_j}{r}\left(\delta_{ij} + \frac{x_i x_j}{r^2}\right) + \frac{1}{4}a^3 V_j \left(\delta_{ij} - 3\frac{x_i x_j}{r^2}\right) \tag{38}$$

This is a combination of a Stokeslet and a potential dipole. With the chosen value of σ_M, the ratio of the coefficient of the potential dipole from FCM to the exact result is $3/\pi$. In general, when using multipole terms for Stokes flow, one must explicitly add the potential dipole. Here it arises naturally and is given to a good approximation given that this term decays quickly with r/a. In the same way, (29) gives the Faxen correction for motion in a nonuniform flow.

Figure 11 shows the streamlines of the FCM result (34) together with the profile of the streamwise velocity component u_1 against x_1. This shows the close match of u_1 to the exact flow at distances $r/a > 1.5$ and the internal flow inside the sphere that resembles a spherical vortex. Indeed the vorticity is smoothed out in the FCM representation and extends into the sphere volume. Similar results for a stresslet flow or a rotlet flow generated by a torque **T** show good agreement with exact results for a single particle with the chosen values of σ_D (Lomholt and Maxey 2003).

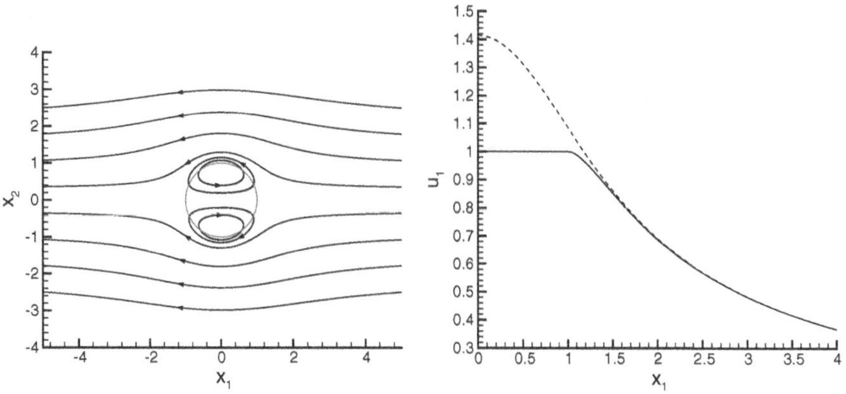

Fig. 11 *Left* Streamlines for FCM Stokes flow past an isolated sphere, shown as the *red circle*, in a uniform flow; *Right* Corresponding velocity profile $u_1(x_1, 0, 0)$ for FCM (*dashed line*) and exact result (*solid line*). Reproduced from Maxey and Patel (2001) with permission (color figure online)

If we consider two spheres of equal size falling under gravity, with equal forces on each, then in Stokes flow they will fall at the same speed and their relative position does not change. Indeed they will fall faster together, with a larger fall speed if they are aligned vertically as opposed to horizontally. The FCM estimates of these fall speeds agree well with the exact results (Lomholt and Maxey 2003). If the spheres are moving in opposing directions, due to equal and opposite forces, then the results lose accuracy if $r/a < 1.25 - 1.5$ depending on the alignment of the particles. For this we need to use correction terms, including the substantial viscous lubrication forces generated by spheres moving with only a narrow gap between them (Dance and Maxey 2003; Yeo and Maxey 2010a).

4.2 Finite Reyonolds Number Flows

A first question is whether the parameterizations for σ_M and σ_D chosen to match the results of Stokes flow are still appropriate for finite Reynolds number flows. In order to address this, we can examine the result for an Oseen flow corresponding to the steady motion of a sphere subject to an applied force $\mathbf{F}$. The flow is steady in the frame of reference moving with the particle and we pose this as a steady, uniform flow $-\mathbf{V}$ past a fixed sphere located at $\mathbf{x} = 0$. We would expect that the corresponding force monopole strength $\mathbf{F}$ would match the standard result for the drag force (10). In this frame of reference, the Oseen flow is $\mathbf{u}'$. For the FCM problem, we rewrite this as $\mathbf{u}' = -\mathbf{V} + \mathbf{u}$ and specify the value of $\mathbf{F}$,

$$-\rho\mathbf{V} \cdot \nabla\mathbf{u} = -\nabla p + \mu\nabla^2\mathbf{u} + \mathbf{F}\Delta_M(\mathbf{x}). \quad (39)$$

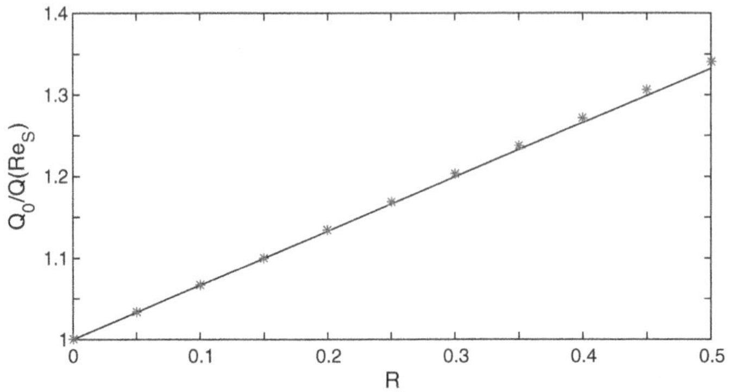

Fig. 12 Comparison of the analytic Oseen drag law (*line*), with the FCM estimate (*stars*) assuming $a/\sigma = \sqrt{\pi}$, plotted as a function of Re_S

Then without any assumptions on a/σ we can evaluate the estimated particle velocity from the solution of this equation and from (29) as $\tilde{V} = FQ(\mathrm{Re_S})/\mu\sigma$, where $\mathrm{Re_S} = \sigma V/\nu$. The question is then what force $\mathbf{F}$ is needed for $\tilde{V}$ to match V. For Stokes flow $Q = Q_0$ and with $a/\sigma = \sqrt{\pi}$, Q_0 matches the Stokes drag law. If this extends to low Reynolds numbers in the Oseen range, then we should find that

$$\frac{Q_0}{Q(\mathrm{Re}_S)} = \frac{V_S}{\tilde{V}} = 1 + \tfrac{3}{8}\mathrm{Re}_S\sqrt{\pi} \tag{40}$$

Figure 12 shows the comparison. There is indeed a good match for low Reynolds numbers. The value of $\mathrm{Re}_S = 0.3$ corresponds to $\mathrm{Re_P} = 1.06$, which may be seen as an upper value for the validity of the Oseen drag law anyway. Details of the calculation are given by Dent (1999). The conclusion is then that the same value of σ_M can be used consistently at all relevant Reynolds numbers. The result also indicates that FCM will yield the same results as the asymptotic procedures for estimating lift and drag forces at low but finite Reynolds numbers discussed in Sect. 2.4. FCM provides a continuous matching of the inner Stokes flow with the outer, finite Reynolds number conditions.

More generally at finite Reynolds numbers we must rely on numerical tests to establish the range of $\mathrm{Re_P}$ and $\mathrm{Re_\Gamma}$ for which the method is accurate. A number of such results are given by Dent (1999) and with further results given by Liu (2004). The conclusion is that FCM is accurate for $\mathrm{Re_P} < 5$, and reasonably accurate for $\mathrm{Re}_P < 10$. We consider a couple of examples.

The first is a particle settling in steady motion along the centerline of a channel at $\mathrm{Re_P}$ of 5. At finite Reynolds numbers, inertial forces will cause the particle to drift away from the wall and fall along the centerline. This contrasts with Stokes flow, where the distance of the particle from the walls remains constant. Figure 13, from Liu et al. (2002), shows the profile for the streamwise velocity component u_1 across the width of the channel, $-5 < x_2/a < 5$ at various locations x_1 relative to

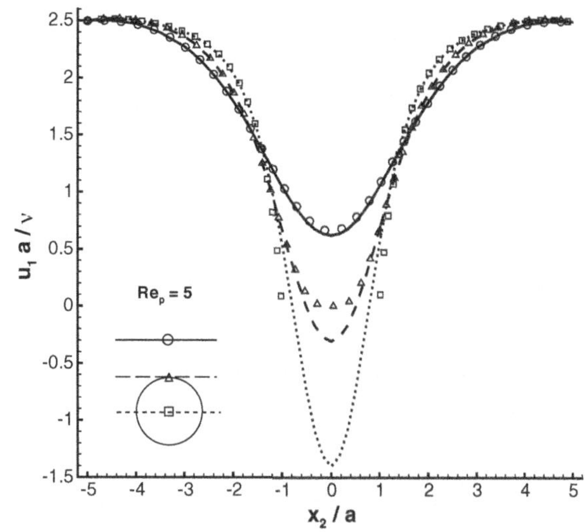

Fig. 13 Profiles of the streamwise velocity of the flow past a sphere settling along the centerline of a channel, comparing FCM and DNS results at various locations. As shown the sphere is fixed and the walls are moving upwards, $Re_P = 5$. Reproduced from (Liu et al. 2002) with permission

the center of the sphere. Periodic boundary conditions are applied at $x_1 = 0, 30a$ and $x_3 = -15a, 15a$. The FCM results are compared to those of a resolved computation of the flow based on the *NEKTAR* spectral/hp element code (SEM). At $x_1 = -2a$ in the wake of the sphere, there is very good agreement of the two results. Inside the particle volume there is the expected overshoot in the FCM velocity but the two simulation results match at about $0.25a$–$0.5a$ away from the sphere surface for $x_1 = 0, -a$. In the SEM computation, the sphere is fixed and the channel walls move vertically upwards with the resulting FCM velocity $-V_1$, so we compare the FCM monopole $\mathbf{F}$ to the drag force $\mathbf{F}_D$ from SEM. This gives $F/\mu\nu = 67.0$ compared to $F_D/\mu\nu = 66.03$. The settling velocity is 70% of that for a freely falling particle under the same force.

At finite Reynolds numbers a stresslet term G_{ij}^S is generated even though none may arise for Stokes flow. This serves to keep the FCM-flow centered on the particle and improves the accuracy of the flow representation. A stresslet is also generated at zero and finite Reynolds numbers if the particle is moving near a wall boundary or another particle. For a pair of identical, freely falling particles at finite Reynolds numbers the effects of fluid inertia cause the drafting, kissing and tumbling as noted by Fortes et al. (1987). The particle trailing in the wake of the leading particle falls more rapidly and catches up (kisses) before the pair then tumbles and separates. The exact sequence depends on the initial horizontal separation of the pair. This phenomena is seen too in the FCM simulations and has been compared with experiments (Lomholt et al. 2002).

Another useful example is to examine the forces on two spherical particles that are fixed in a Poiseuille flow as shown in Fig. 14. The flow has periodic boundary conditions at $x_1 = 0, 20a$ and $x_3 = -3.5a, 3.5a$, with planar walls at $x_2 = -3.5a, 3.5a$. The two particles are coplanar and centered at $x_2 = 1.5$ with a streamwise separation

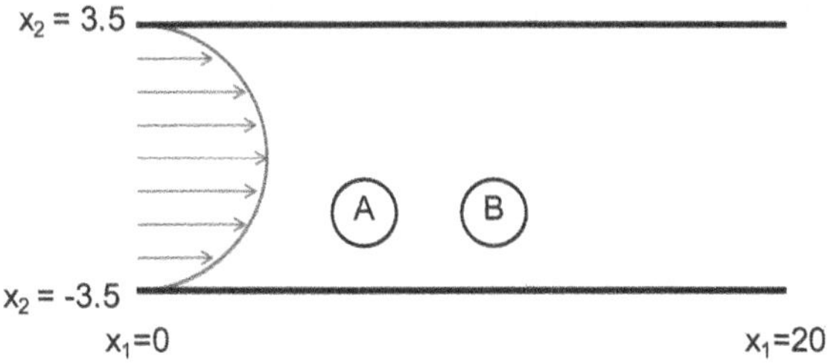

Fig. 14 Two particles held fixed in a Poiseuille flow within a planar channel, channel height $7a$, particles centered at $2a$ from lower wall, separated by $4a$. Reynolds number based on centerline velocity is 2.45

$4a$. Again we can compare FCM results with SEM, using a penalty method to keep the FCM particles fixed (Liu 2004). Based on the Poiseuille flow velocity u_0 at the particle centers, $Re_P = 2$. The drag force on the upstream particle is 5% larger than on the second, consistent with the wake effect. There is a lift force on the upstream particle $F_L/\mu a u_0 = -1.33$ directed towards the centerline as one may expect. The FCM estimate is -1.36. On the downstream particle the lift force is instead towards the wall $F_L/\mu a u_0 = 0.43$; the FCM estimate is 0.41. This shows how the disturbance flow of one particle can modify the motion of another and that estimates of drag and lift, or equations of motion such as (6), for isolated particles need careful interpretation. Even approximate methods such as FCM can help in understanding the interactions of particle motion.

5 Applications to Suspension Flows

5.1 Particle Settling

Here we consider a few examples of FCM applied to the motion of freely suspended particles. The first is the settling of randomly dispersed particles under gravity in a homogeneous suspension. This is a topic with a long history. The theoretical problem consists of a large number of freely suspended particles, of the same radius a, that are denser than the fluid and so experience a net negative buoyancy force due to gravity. The fluid is otherwise at rest and the particles start to settle from their initial, random uniform distribution characterized by the volume fraction c. The first question is how quickly do the particles settle on average $\langle V \rangle$ and how does this compare to the terminal fall speed W_S of an isolated particle.

For a numerical simulation, we set up a periodic domain to model a homogeneous system with the length L of the periodic box large compared to a. The FCM equation (26) is computed for either Stokes flow or finite low Reynolds number conditions using a Fourier pseudo-spectral method as described by Climent and Maxey (2003). First for Stokes flow, particle and fluid inertia are negligible, and we can compute the resulting motion using both the force monopole and dipole terms. As noted earlier, a contact force is needed to represent the short-range effects of solid body contacts where small roughness irregularities on the particle surface prevent direct contact (Rampall et al. 1997). This preserves the void fraction of the suspension and prevents the overlap of particles. Often the specific form of the contact force model is not critical and we typically use a model that goes smoothly to zero for $|\mathbf{Y}^{(A)} - \mathbf{Y}^{(B)}|$ larger than R_{ref} but rises sharply as the particles approach contact (Dance et al. 2004). For this type of flow with the settling of identical particles at lower volume fractions $c < 20\%$, viscous lubrication forces are not a significant factor.

The first condition to impose is that there is a counter-balancing uniform pressure gradient on the fluid to support the net weight of the suspended particles. In a Fourier scheme, this is imposed for the zero wavenumber component of the flow $\hat{\mathbf{u}}(\mathbf{k}, t)$ and $\mathbf{k} = 0$. Without this condition there would be no stationary equilibrium. Batchelor (1972) pointed out that while pairwise hydrodynamic interactions between the particles may tend to increase the settling rate, their motion also creates a net updraft of fluid to counter the volume displacement of the falling particles. Overall, there is hindered settling and $\langle V \rangle < W_S$. This condition of zero total volume flux corresponds to setting $\hat{\mathbf{u}}(0, t) = 0$. Physically, it is supposed that there is some lower boundary to which the particles would settle and stop and which would also block any flow. This is the standard frame of reference in which sedimentation is considered. Any non-zero value for $\hat{\mathbf{u}}(0, t)$ would simply be a constant added to both the fluid and particle velocities, which would not alter the relative motion from settling.

Figure 15 shows the simulation results for $\langle V \rangle / V_0$ for Stokes flow as compared to the experimental measurements of Nicolai et al. (1995). The value of V_0 is the fall speed of a single particle in the periodic domain and differs slightly from W_S. Batchelor (1972) estimated this ratio for the hindered settling velocity to be $(1 - 6.55c + O(c^2))$. The curve shown is based on a standard Richardson–Zaki correlation of the data fitted to $\langle V \rangle / V_0 = (1 - c)^{4.5}$. The simulation results are based on a box $L = 128\Delta x$ and $\sigma_M = 1.5\Delta x$ so that $L/a = 48$ approximately. Tests were made that showed that the stresslet terms did not have much influence on the

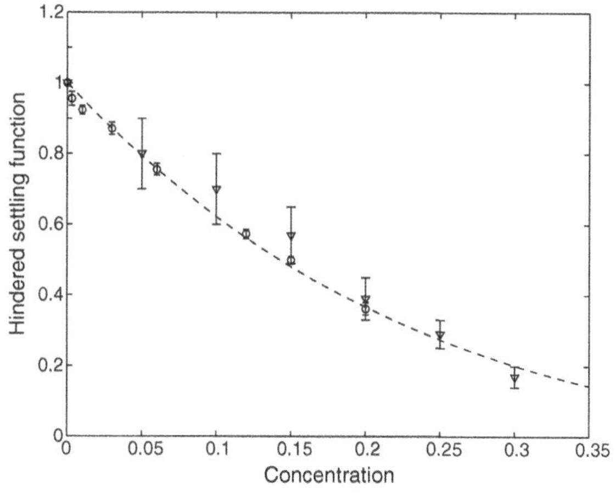

Fig. 15 Average settling rate of particles in a Stokes suspension: FCM results (*circles*), experiments of Nicolai et al. (1995) (*triangles*), *dashed line* shows standard a Richardson–Zaki correlation, $(1 - c)^{4.5}$. Reproduced from Climent and Maxey (2003) with permission

statistical results and these were omitted from the simulations (Climent and Maxey 2003).

There is a discussion of settling in a homogeneous suspension for Stokes flow in the review papers of Davis and Acrivos (1985), Davis (1996) and in Guazzelli and Morris (2012). Recent results are summarized by Guazzelli and Hinch (2011). The variation of $\langle V\rangle < W_S$ with c is a robust result with a Richardson–Zaki exponent usually about 5 and is not sensitive to the domain size L/a. However it was pointed out by Caflisch and Luke (1985) that this idealized problem would lead to a divergence in the level of particle velocity fluctuations V', where $V_j' = \langle(V_j - \langle V_j\rangle)^2\rangle^{1/2}$, with increasing box size. An explanation for this is given by Hinch (1988) in terms of the formation of large clusters that settle collectively and so give $V'/W_S \sim (cL/a)^{1/2}$. This feature is seen too in the FCM simulations as shown in Fig. 16 for $c = 0.06$ as L/a varies from 16 to 96. The circles in the figures are results for Stokes flow and the figure also verifies the lack of variation of $\langle V\rangle/V_0$ with domain size. The increase in V' with L/a is consistent with Hinch's estimate.

The divergence of V' is not seen in experiments and it has been discussed how the experiments in a container with a sharp sedimentation front that slowly descends will differ from the idealized problem simulated numerically (Guazzelli and Hinch 2011). The aim of the FCM study (Climent and Maxey 2003) was to see how finite Reynolds number effects would modify the flow. A feature of Stokes flow is the long-range effect of a Stokeslet flow which decays relatively slowly with distance and can

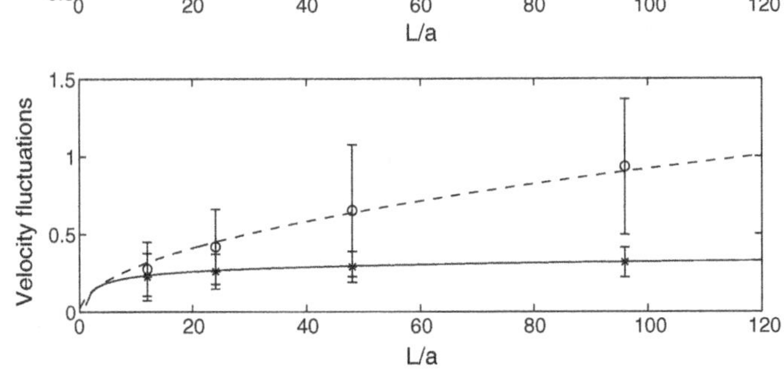

Fig. 16 Variations of mean settling velocity (*upper*) and rms velocity fluctuations (*lower*), scaled by V_0, with domain size L/a. Results are for Stokes flow (*circles*) and for $Re_P = 5$ (*stars*) at $c = 0.06$; statistical error bars are shown for these time averages. Also shown are the scalings of Caflisch and Luke (1985) for Stokes flow (*dashed line*) and Koch (1993) for $Re_P = 5$ (*solid line*). Reproduced from Climent and Maxey (2003) with permission

contribute to cluster formation or long-range correlations in the velocities. Koch (1993) and later Yin and Koch (2007) had noted that the long-range effects of a point force in Oseen flow, an Oseenlet, would be much weaker. The FCM simulations in principle should account for the particle as well as the fluid inertia through (35). Tests verified that for $0.9 < m_P/m_F < 1.5$, the particle acceleration is small compared to gravitational acceleration g and so a simple constant external force is sufficient to represent the dynamics. The particle Reynolds number Re_P was varied over 0.1, 1, 5 and 10.

Figure 16 shows the results for $\mathrm{Re}_P = 5$ and $c = 0.06$ as L/a is varied, where $\mathrm{Re}_P = 2aV_0/\nu$. The mean settling rate $\langle V \rangle / V_0$ is less at higher Re_P and again does not vary with domain size. The velocity fluctuations are less too but do not increase with domain size as they did for Stokes flow. Koch (1993) had indicated that an Oseen flow model would predict $V'/V_0 \sim \{\ln(L/a)\}^{1/2}$ and this is shown in comparison with the data. There may still be an increase with domain size but it is weak.

The effect of fluid inertia can also be seen by comparing the vertical sections of the instantaneous flow, as shown in Fig. 17 where $c = 0.06$ and $L/a = 48$. Visually there is more tendency for clusters of particles to form in Stokes flow with larger spacing between them, while at finite Re the particles are more evenly dispersed. This would indicate less tendency for large scale overturning to develop. The drafting–kissing–tumbling of particles in finite Re flows means that even through pairwise interactions the microstructure is continually evolving, while in Stokes flow three-way particle interactions are needed to change the relative configuration of the particles. Further data shows that the average minimum distance between particles increases with Re and the integral length scale for the fluid velocity fluctuations is shorter.

5.2 Couette Flow

A second example is shear flow in a suspension of neutrally buoyant particles at zero Reynolds number and volume fractions of $c = 0.2 - 0.4$. In this context we must pay close attention to the viscous lubrication forces between particles that are nearly in contact. In a shear flow, particles are often moving in opposite directions either towards each other or away and many simulation methods, including FCM, need to make corrections for this. Figure 18 illustrates the forces generated by two particles in relative motion when the gap between them ϵa is small and below the resolution of the simulation method. The estimates of lubrication forces, see Kim and Karrila (1991), are:

$$
\begin{aligned}
F_1 &\propto \mu a \left(V_1^A - V_1^B\right) \frac{1}{\varepsilon} \\
F_2 &\propto \mu a \left(V_2^A - V_2^B\right) \log \varepsilon
\end{aligned}
\tag{41}
$$

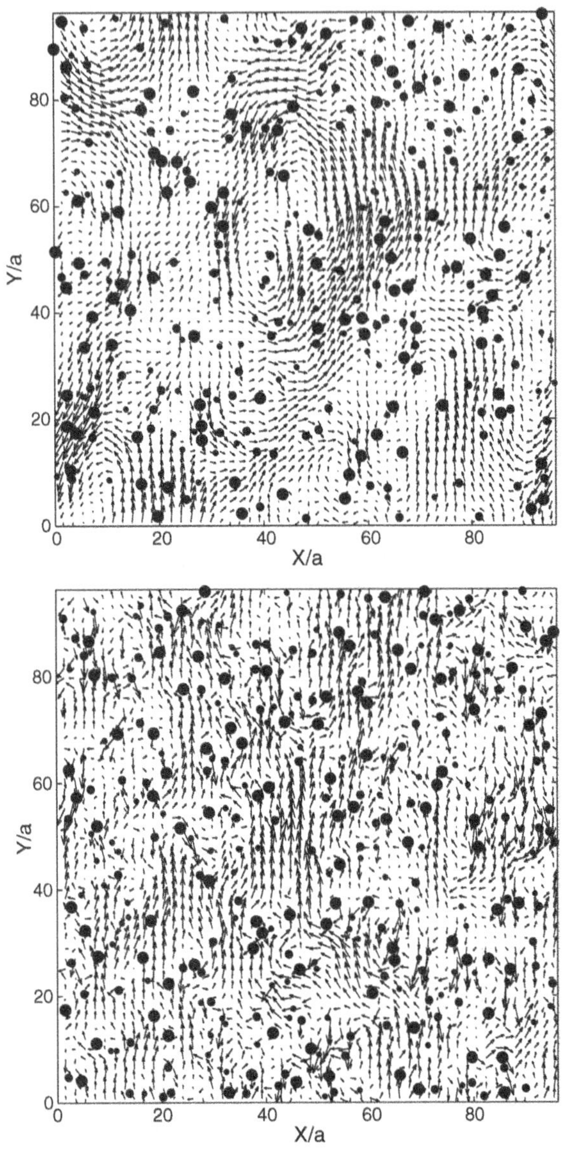

Fig. 17 Sample vertical slices for Stokes flow (*upper*) and $\mathrm{Re}_P = 5$ (*lower*) for the settling particles at $c = 0.06$. The particles are shown by their area fraction intersecting the plane together with local fluid velocity. Reproduced from Climent and Maxey (2003) with permission

This is incorporated within FCM by forming $\mathbf{F}_{Lub} = \mathbf{F}_{FCM} + (\mathbf{F}_{Lub} - \mathbf{F}_{FCM})$. This blends the features captured by FCM with the analytical results for pairwise interactions of particles. There are further terms for lubrication forces and torques as described by Dance and Maxey (2003), Yeo and Maxey (2010a).

Guazzelli and Morris (2012) provide a good overview of the dynamics of shear flow suspensions of non-Brownian particles. In the dilute limit, the primary effect

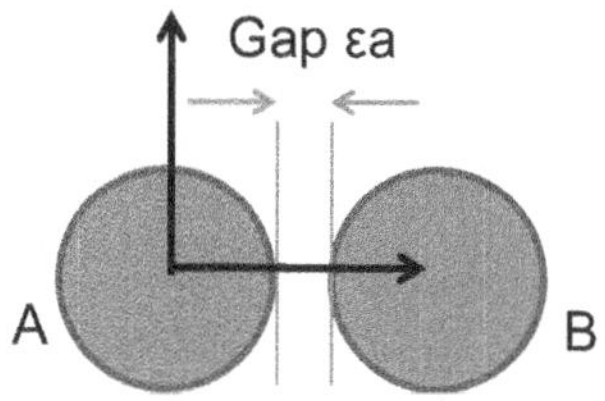

Fig. 18 Sketch of the lubrication forces between two particles, with forces on A due to the presence of B and gap width ϵa

for uniform shear flow is an enhanced effective shear viscosity which according to the Einstein estimate is $\mu_{eff} = \mu(1 + 2.5c)$. Within the context of FCM, an isolated sphere subject to an external flow with rate of strain E_{ij}^{∞} will generate a stresslet $G_{ij}^{S} = (20/3)\pi\mu a^3 E_{ij}^{\infty}$. A distribution of stresslets from a collection of well separated particles will generate a body force density in (26) that can be written as

$$f_i = \frac{\partial}{\partial x_j} \sum_n \Delta_D(\mathbf{x} - \mathbf{Y}^n) G_{ij}^{S} \tag{42}$$

This has the form of a divergence of a stress tensor associated with the particle phase and

$$\frac{4}{3}\pi a^3 \sum_n \Delta_D(\mathbf{x} - \mathbf{Y}^n)$$

is an effective local volume fraction for the particle phase. An ensemble average of these results yields the Einstein estimate for particle-induced increase in shear viscosity $2.5\mu c$. Overall, the particles lead to an enhanced viscous dissipation in the flow.

As the volume fraction of particles increases, two-particle interactions are important and so too are lubrication forces. A uniform shear flow is the superposition a solid body rotation and a straining flow with principal axes at 45° to the flow. In the compressive direction, with negative rate of strain, the particles are squeezed together by the flow and then separated in the extensive direction. Both will generate significant particle stresses, including normal stresses, through these lubrication forces. It is clear then that the relative configuration of the particles is important. This microstructure is often characterized by the pair distribution function $g(\mathbf{r})$, which is proportional to the probability density for the center of a particle being located at a point $\mathbf{r}$ relative to the center of a reference particle. The dynamics and microstructure of suspensions in uniform shear flow have been studied extensively with Stokesian dynamics (SD) as a simulation method (Sierou and Brady 2002).

The presence of channel walls in a Couette flow can significantly alter the relative distribution of the particles within a suspension even if the core region resembles a uniform homogeneous shear. This is illustrated by FCM simulations of Couette flow in which both the volume fraction of the particles and the channel width H/a are varied (Yeo and Maxey 2010b). The simulations are set up with periodic boundary conditions in the streamwise direction, $L_x = 30a$, and in the spanwise direction,

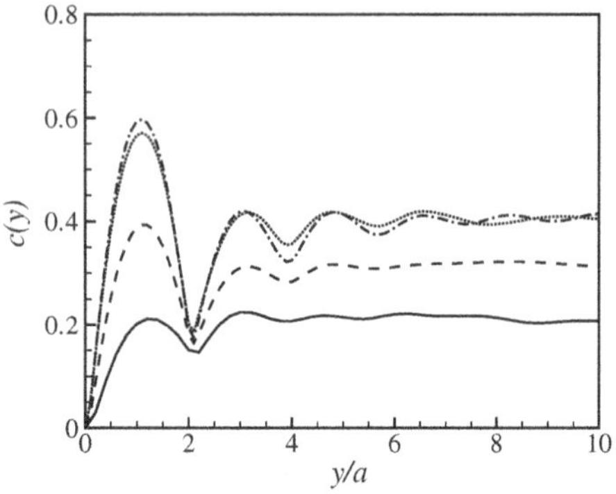

Fig. 19 Particle void fraction profiles in Couette flow, $H = 20a$, for bulk concentrations: $c = 0.2$ (*solid*), $c = 0.3$ (*dashed*), $c = 0.4$ (*dash-dot*). Also shown is $c = 0.4$ (*dotted*) with smaller $\epsilon_{\min}$. Reproduced from Yeo and Maxey (2010b) with permission

$L_z = 20a$. In the wall normal direction, $0 \le y \le H$, a spectral element discretization is used. The paper gives details on the implementation of the no-slip boundary conditions at the wall in conjunction with FCM and the lubrication forces between the wall and the particles. A short-range contact barrier is also implemented at the wall, similar to that between particles.

Figure 19 shows profiles of the local void fraction near the lower wall for bulk volume fractions of 20, 30 and 40% for a channel of width $H = 20a$. The profile is calculated as the fraction of a planar slice, at a chosen y/a, that is intersected by particles. When ensemble averaged it indicates the likelihood of finding a particle located relative to the wall. There is an excluded volume effect and a strong first peak forms at $y/a = 1$ corresponding to a layer of particles adjacent to the wall. There is a local minimum at $y/a = 2$ followed by secondary peaks at $y/a = 3$ and $y/a = 5$. These are progressively weaker as the layering becomes less coherent. This is a feature of flow with a smooth wall and monodisperse particles which are large enough that Brownian motion is not a factor. Generally, the value of R_{ref} is set so that the minimum gap between particles corresponds to $\epsilon_{min} = 0.005$. An additional result with $\epsilon_{min} = 0.0005$ indicates that this feature is not sensitive to the choice of R_{ref}.

Figure 20 shows that these near-wall features are similar if the channel width H/a is varied. The layering is more pronounced at higher bulk volume fractions.

We can also look at the pair distribution function between particles when we choose the reference particle to be in the core of the channel or nearer to a wall. Figure 21 gives a contour plot of $g(\mathbf{r}; Y_2)$ first when the location of the reference particle is at $8 < Y_2/a < 12$ close to the center of the channel. This is taken in a vertical slice through the center of the reference particle at $c = 0.4$. There is an exclusion zone $r < 2a$ in which $g = 0$ and region just beyond this where the probability of finding a second particle is high. This ring is concentrated about the compressive principal axes for the shear flow. The segments in the extensive directions have low probability. If the reference particle is closer to the lower wall in a region centered at $Y_2 = 6a$, then there is a clear influence of the wall on the lower side. The upper side has less influence from the wall but the ring-like profile is still modified. Together

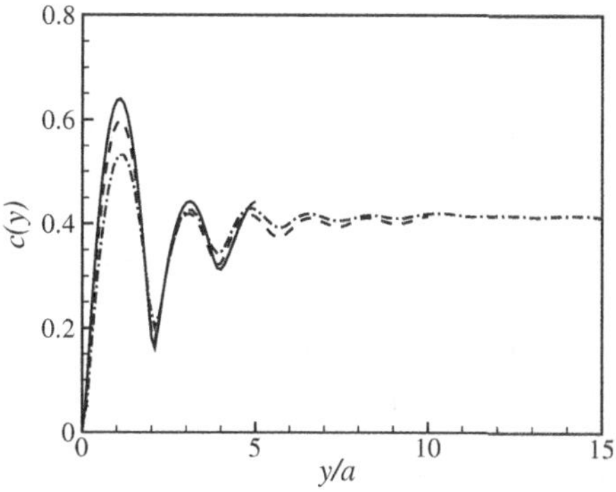

Fig. 20 Near-wall particle void fraction profiles in Couette flow for bulk concentration $c = 0.4$, channel heights: $H/a = 10$ (*solid*), $H/a = 20$ (*dashed*) and $H/a = 30$ (*dash-dot*). Reproduced from Yeo and Maxey (2010b) with permission

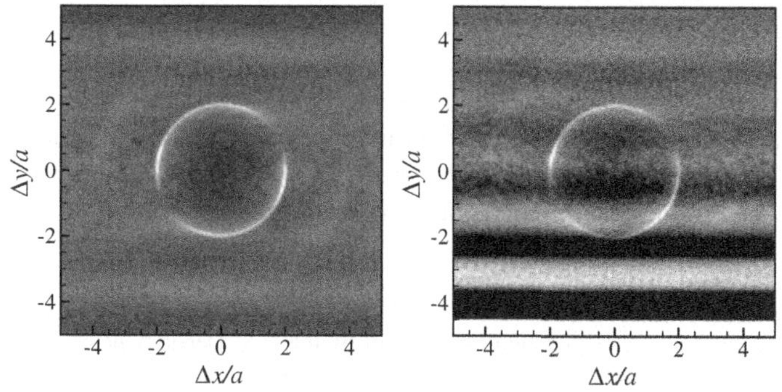

Fig. 21 Comparison of the pair distribution functions in the core of the channel, for particles centered $8 < y/a < 12$, and in the wall region, for particles centered $5.5 < y/a < 6.5$. Bulk concentration $c = 0.4$, $H = 20a$. Light contours represent high probability. Reproduced from Yeo and Maxey (2010b) with permission

the results suggest that near-wall dynamics may extend out to $Y_2/a = 7$ or so. Yeo and Maxey (2010b) provide additional details about the particle stresses including the normal stresses.

In a further study, Yeo and Maxey (2010c) explore the Couette flow at higher volume fractions in the dense suspension range $0.48 < c < 0.6$. The ordering effect created by the walls is even more pronounced and can extend the full width of the channel. In fact a form of crystallization can occur depending on whether H is commensurate with the particle diameter. If an ordered set of layers form the effective shear viscosity is reduced, while if not it is substantially higher.

6 Comments

The force coupling method has been used to investigate other suspension flows including a pressure-driven Poiseuille flow (Yeo and Maxey 2011) and homogeneous shear flow at finite Reynolds numbers (Yeo and Maxey 2013). These references also note additional modifications of the method to speed up the computations or extend the range of particle to fluid densities that can be covered for inertial flows. FCM may also be applied to ellipsoidal particles at both zero and low, finite Reynolds numbers (Liu et al. 2009). This simply requires a generalization of (27) to reflect an ellipsoid instead of a sphere but no other modifications are needed. This works well for moderate aspect ratios of a spheroid but would not be efficient for an elongated shape or fiber. FCM has also been used to investigate the inertial migration of neutrally buoyant particles in a laminar flow within a channel of square section, relevant to the inertial focusing of particles (Abbas et al. 2014). The discussion has focused on FCM as a simulation method, but we have pointed to the way in which it provides a natural framework for discussing particle stresses or for simplified dynamic models.

In the end there are many ways to model or numerically simulate particles suspended in a flow. Each method has it specific range of application or set of restrictions. Taken together with experimental measurements, simulations present a more compete picture of the dynamics. Reduced or simplified simulation methods are especially helpful in isolating the essential dynamics.

Acknowledgements We would like to thank Dr. Kyongmin Yeo, now at IBM Watson Research Center for his help with the numerical simulations. Support through an award CBET-1133106 from the U.S. National Science Foundation and support by the U.S. Department of Energy Office of Science, Office of Advanced Scientific Computing Research, Applied Mathematics program as part of the Colloboratory on Mathematics for Mesoscopic Modeling of Materials (CM4), under award number DE-SC0009247 are both gratefully acknowledged.

References

M. Abbas, P. Magaud, Y. Gao, and S. Geoffroy. Migration of finite sized particles in a laminar square channel flow from low to high Reynolds numbers. *Phys. Fluids*, 26:123301, 2014.

R.J. Adrian and J. Westerweel. *Particle Image Velocimetry*. Number 30 in Cambridge Aerospace Series. Cambridge University Press, 2011.

E.S. Asmolov. The inertial lift on a spherical particle in a plane Poiseuille flow at large channel Reynolds number. *J. Fluid Mech.*, 381:63–87, 1999.

T.R. Auton, J.C.R. Hunt, and M. Prudhomme. The force exerted on a body in inviscid steady non-uniform rotational flow. *J. Fluid Mech.*, 197:241–257, 1988.

P. Bagchi and S. Balachandar. The force exerted on a body in inviscid steady non-uniform rotational flow. *Phys. Fluids*, 14:2719–2737, 2002.

P. Bagchi and S. Balachandar. Inertial and viscous forces on a rigid sphere in straining flows at moderate Reynolds numbers. *J. Fluid Mech.*, 481:105–148, 2003.

S. Balachandar and John K. Eaton. Turbulent dispersed multiphase flow. Annu. Rev. Fluid Mech., 42:111–133, 2010.

A.B. Basset. On the motion of a sphere in a viscous liquid. *Phil. Trans. Royal Soc. London. Series A*, 179:43–63, 1888.
A.B. Basset. Descent of a sphere in a viscous fluid. *Quarterly Journal of Pure and Applied Mathematics*, 41:369–381, 1910.
G.K. Batchelor. *An Introduction to Fluid Dynamics*. Cambridge University Press, 1967.
G.K. Batchelor. Sedimentation in a dilute dispersion of spheres. *J. Fluid Mech.*, 52(02):245–268, 1972.
L. Bergougnoux, G. Bouchet, D. Lopez, and É. Guazzelli. The motion of solid spherical particles falling in a cellular flow field at low Stokes number. *Phys. Fluids*, 26(9):093302, 2014.
W. P. Breugem. A second-order accurate immersed boundary method for fully resolved simulations of particle-laden flows. *J. Comput. Physics*, 231:4469–4498, 2012.
R.E. Caflisch and J.H.C. Luke. Variance in the sedimentation speed of a suspension. *Phys. Fluids*, 28(3):759–760, 1985.
F. Candelier, J.R. Angilella, and M. Souhar. On the effect of the Boussinesq-Basset force on the radial migration of a Stokes particle in a vortex. *Phys. Fluids*, 16:1765–1776, 2004.
E. J. Chang and M. R. Maxey. Unsteady flow about a sphere at low to moderate Reynolds number. Part 1. Oscillatory motion. *J. Fluid Mech.*, 277:347–379, 1994.
E. J. Chang and M. R. Maxey. Unsteady flow about a sphere at low to moderate Reynolds number. Part 2. Accelerated motion. *J. Fluid Mech.*, 303:133–153, 1995.
E. Climent and M.R. Maxey. Numerical simulations of random suspensions at finite Reynolds numbers. *Int. J. Multiphase Flow*, 29:579–601, 2003.
S.L. Dance and M.R. Maxey. Incorporation of lubrication effects into the force-coupling method for particulate two-phase flow. *J. Comput. Phys.*, 189:212–238, 2003.
S.L. Dance, E. Climent, and M.R. Maxey. Collision barrier effects on the bulk flow in a random suspension. *Phys. Fluids*, 16:828–831, 2004.
R.H. Davis. Hydrodynamic diffusion of suspended particles: a symposium. *J. Fluid Mech.*, 310:325–335, 1996.
R.H. Davis and A. Acrivos. Sedimentation of noncolloidal particles at low Reynolds numbers. *Annu. Rev. Fluid Mech.*, 17:91–118, 1985.
G.L. Dent. *Aspects of particle sedimentation in dilute flows at finite Reynolds numbers*. PhD thesis, Brown University, 1999.
D. Di Carlo. Inertial microfluidics. *Lab on a Chip*, 9:3038–3046, 2009.
P.G. Drazin and W.H. Reid. *Hydrodynamic Stability, 2nd ed.* Cambridge University Press, 2004.
A.F. Fortes, D.D. Joseph, and T.S. Lundgren. Nonlinear mechanics of fluidization of beds of spherical particles. *J. Fluid Mechanics*, 177:467–483, 1987.
R. Gatignol. The Faxén formulas for a rigid particle in an unsteady non-uniform Stokes-flow. *Journal de Mécanique Théorique et Appliquée*, 1(2):143–160, 1983.
R. Glowinski, T.-W. Pan, T.I. Hesla, and D.D. Joseph. A distributed Lagrange multiplier/fictitious domain method for particulate flows. *Int. J. Multiphase Flows*, 25:755–794, 1999.
W.W. Grabowski and L.-P. Wang. Growth of cloud droplets in a turbulent environment. *Annu. Rev. Fluid Mech.*, 45:293–324, 2013.
É. Guazzelli and J. Hinch. Fluctuations and instability in sedimentation. *Annu. Rev. Fluid Mech.*, 43:97–116, 2011.
É. Guazzelli and J.F. Morris. *A Physical Introduction to Suspension Dynamics*. Cambridge University Press, New York, 2012.
E.J. Hinch. Sedimentation of small particles. In E. Guyon, J.-P. Nadal, and Y. Pomeau, editors, *Disorder and Mixing*, pages 153–161. Kluwer Academic, 1988.
H.H. Hu, N.A. Patankar, and M.Y. Zhu. Direct numerical simulations of fluid-solid systems using the arbitrary Lagrangian-Eulerian technique. *J. Comput. Phys.*, 169:427–462, 2001.
R.H.A. Ijzermans, E. Meneguz, and M.W. Reeks. Segregation of particles in incompressible random flows: singularities, intermittency and random uncorrelated motion. *J. Fluid Mech.*, 653:99–136, 2010.

G.B. Jeffery. The motion of ellipsoidal particles immersed in a viscous fluid. *Proc. Royal Soc. London. Series A*, 102:161–179, 1922.
M. Jenny, J. Dušek, and G. Bouchet. Instabilities and transition of a sphere falling or ascending freely in a Newtonian fluid. *J. Fluid Mech.*, 508:201–239, 2004.
D.H. Kelley and N.T. Ouellette. Onset of three-dimensionality in electromagnetically driven thin-layer flows. *Phys. Fluids*, 23(4):045103, 2011.
S. Kim and S.J. Karrila. *Microhydrodynamics: Principles and Selected Applications*. Butterworth-Heinemann, Boston, 1991.
D.L. Koch. Hydrodynamic diffusion in dilute sedimenting suspensions at moderate Reynolds numbers. *Phys. Fluids A*, 5:1141–1155, 1993.
C.J. Lawrence and R. Mei. Long-time behaviour of the drag on a body in impulsive motion. *J. Fluid Mech.*, 283:307–327, 1995.
Y. Ling, M. Parmar, and S. Balachandar. A scaling analysis of added-mass and history forces and their coupling in dispersed multiphase flows. *Int. J. Multiphase Flow*, 57:102–114, 2013.
D. Liu. *Spectral Element/Force Coupling Method: Application to Colloidal Micro-Devices and Self-Assembled Particle Structures in 3D Domains*. PhD thesis, Brown University, 2004.
D. Liu, M.R. Maxey, and G.E. Karniadakis. A fast method for particulate microflows. *J. Microelectromechanical Systems*, 11:691–702, 2002.
D. Liu, E.E. Keaveny, M.R. Maxey, and G.E. Karniadakis. Force-coupling method for flows with ellipsoidal particles. *J. Comput. Phys*, 228:3559–3581, 2009.
S. Lomholt and M.R. Maxey. Force-coupling method for particles sedimenting in a channel: Stokes flow. *J. Comput. Phys.*, 184:381–405, 2003.
S. Lomholt, B. Stenum, and M.R. Maxey. Experimental verification of the force coupling method for particulate flows. *Int. J. Multiphase Flow*, 28:225–246, 2002.
E. Loth and A. J. Dorgan. An equation of motion for particles of finite Reynolds number and size. *Envir. Fluid Mech.*, 9:187–206, 2009.
P.M. Lovalenti and J.F. Brady. The hydrodynamic force on a rigid particle undergoing arbitrary time-dependent motion at small Reynolds number. *J. Fluid Mech.*, 256:561–605, 1993.
P.M. Lovalenti and J.F. Brady. The temporal behaviour of the hydrodynamic force on a body in response to an abrupt change in velocity at small but finite Reynolds number. *J. Fluid Mech.*, 293:35–46, 1995.
J. Magnaudet, M. Rivero, and J. Fabre. Accelerated flows past a rigid sphere or a spherical bubble. Part 1. Steady straining flow. *J. Fluid Mech.*, 284:97–135, 1995.
R. Mallier and M. Maxey. The settling of nonspherical particles in a cellular flow field. *Phys. Fluids A*, 3:1481–1494, 1991.
M. R. Maxey and J. J. Riley. Equation of motion for a small rigid sphere in a nonuniform flow. *Phys. Fluids*, 26:883–889, 1983.
M.R. Maxey. The gravitational settling of aerosol particles in homogeneous turbulence and random flow fields. *J. Fluid Mech.*, 174:441–465, 1987a.
M.R. Maxey. The motion of small spherical particles in a cellular flow field. *Phys. Fluids*, 30:1915–1928, 1987b.
M.R. Maxey and S. Corrsin. Gravitational settling of aerosol particles in randomly oriented cellular flow fields. *J. Atmos. Sci.*, 43:1112–1134, 1986.
M.R. Maxey and B.K. Patel. Localized force representations for particles sedimenting in Stokes flow. *Int. J. Multiphase Flow*, 27:1603–1626, 2001.
J.B. McLaughlin. Inertial migration of a small sphere in linear shear flows. *J. Fluid Mech.*, 224:261–274, 1991.
J.B. McLaughlin. The lift on a small sphere in wall-bounded linear shear flows. *J. Fluid Mech.*, 246:249–265, 1993.
R. Mei and R.J. Adrian. Flow past a sphere with an oscillation in the free-stream and unsteady drag at finite Reynolds number. *J. Fluid Mech.*, 237:323–341, 1992.
R. Mei and R.J. Adrian. Flow due to an oscillating sphere and an expression for unsteady drag on the sphere at finite Reynolds number. *J. Fluid Mech.*, 270:133–174, 1994.

E. Meiburg and B. Kneller. Turbidity currents and their deposits. *Annu. Rev. Fluid Mech.*, 42:135–156, 2010.

R Monchaux, M. Bourgoin, and A. Cartellier. Analyzing preferential concentration and clustering of inertial particles in turbulence. *Int. J. Multiphase Flow*, 40:1–18, 2012.

H. Nicolai, B. Herzhaft, E.J. Hinch, L. Oger, and É. Guazzelli. Particle velocity fluctuations and hydrodynamic self-diffusion of sedimenting non-brownian spheres. *Phys. Fluids*, 7(1):12–23, 1995.

C.W. Oseen. Uber die Stokessche formel und uber eine verwandte aufgabe in der hydrodynamik. *Ark. Mat. Astron. Fys.*, 6(29):1–20, 1910.

I. Rampall, J.R. Smart, and D.T. Leighton. The influence of surface roughness on the particle-pair distribution function of dilute suspensions of non-colloidal spheres in simple shear flow. *J. Fluid Mech.*, 339:1–24, 1997.

M Rivero, J Magnaudet, and J Fabre. Quelques résultats nouveaux concernant les forces exercées sur une inclusion sphérique par un écoulement accéléré. *Comptes rendus de l'Académie des sciences. Série 2, Mécanique, Physique, Chimie, Sciences de l'univers, Sciences de la Terre*, 312(13):1499–1506, 1991.

J. Rubin, C.K.R.T. Jones, and M.R. Maxey. Settling and asymptotic motion of aerosol particles in a cellular flow field. *J. Nonlinear Science*, 5(4):337–358, 1995.

S.I. Rubinow and J.B. Keller. The transverse force on a spinning sphere moving in a viscous fluid. *J. Fluid Mech.*, 11(03):447–459, 1961.

P. G. Saffman. The lift on a small sphere in a slow shear flow. *J. Fluid Mech.*, 22:385–400, 1965.

P.G. Saffman. Corrigendum. *J. Fluid Mech.*, 31:624, 1968.

T. Sano. Unsteady flow past a sphere at low Reynolds number. *J. Fluid Mech.*, 112:433–441, 1981.

R.A. Shaw. Particle-turbulence interactions in atmospheric clouds. *Annu. Rev. Fluid Mech.*, 35(1):183–227, 2003.

H. Shin and M.R. Maxey. Chaotic motion of nonspherical particles settling in a cellular flow field. *Phys. Rev. E*, 56:5431–5444, 1997.

A. Sierou and J. F. Brady. Rheology and microstructure in concentrated noncolloidal suspensions. *J. Rheol.*, 46:1031–1056, 2002.

H. Stommel. Trajectories of small bodies sinking slowly through convection cells. *J. Marine Res.*, 8:24–29, 1949.

H.A. Stone. Philip Saffman and viscous flow theory. *J. Fluid Mech.*, 409:165–183, 2000.

G.I. Taylor and A.E. Green. Mechanism of the production of small eddies from large ones. *Proc. Royal Soc. London. Series A*, 158(895):499–521, 1937.

A.G. Tomboulides and S.A. Orszag. Numerical investigation of transitional and weak turbulent flow past a sphere. *J. Fluid Mech.*, 416:45–73, 2000.

M. Uhlmann. An immersed boundary method with direct forcing for the simulation of particulate flows. *J. Comput. Physics*, 209:448–476, 2005.

M. A. T. van Hinsberg, J. H. M. ten Thije Boonkkamp, and H. J. H. Clercx. An efficient, second order method for the approximation of the Basset history force. *J. Comput. Phys.*, 230:1465–1478, 2011.

L.-P. Wang and M.R. Maxey. Settling velocity and concentration distribution of heavy particles in homogeneous isotropic turbulence. *J. Fluid Mech.*, 256:27–68, 1993.

M. Wilkinson, B. Mehlig, S. Östlund, and K.P. Duncan. Unmixing in random flows. *Phys. Fluids*, 19(11):113303, 2007.

K. Yeo and M.R. Maxey. Simulation of concentrated suspensions using the force-coupling method. *J. Comput. Phys.*, 229:2401–2421, 2010a.

K. Yeo and M.R. Maxey. Dynamics of concentrated suspensions of non-colloidal particles in Couette flow. *J. Fluid Mech.*, 649:205–231, 2010b.

K. Yeo and M.R. Maxey. Ordering transition of concentrated suspensions of non-Brownian particles in a confined steady shear flow. *Phys. Rev. E*, 81:051502, 2010c.

K. Yeo and M.R. Maxey. Numerical simulations of concentrated suspensions of monodisperse particles in a Poiseuille flow. *J. Fluid Mech.*, 682:491–518, 2011.

K. Yeo and M.R. Maxey. Dynamics and rheology of concentrated, finite-Reynolds-number suspensions in a homogeneous shear flow. *Phys. Fluids*, 25:053303, 2013.

X. Yin and D.L. Koch. Hindered settling velocity and microstructure in suspensions of solid spheres with moderate Reynolds numbers. *Phys. Fluids*, 19:093302, 2007.

Modeling and Simulation of Finite-Size Particles in Turbulence

Holger Homann

Abstract This course gives an introduction to the dynamics of finite-size particles moving in laminar and especially turbulent flows. Finite-size means that their diameter is larger than the smallest active scale of the carrier flow and that their slipping velocity is not negligible. An emphasis will be on the differences in the dynamics of these particles compared with those that are much smaller than any scale of flow variation. We will see that the differences are substantial: The dynamics of finite-size particles is hard to model, numerical simulations are challenging, large particles have a complicated imprint on the carrier flow and their mutual hydrodynamic interactions show rich properties.

1 Introduction

Many environmental and industrial flows are in a turbulent state. This means that inertial effects of the gas or liquid are important which might be for example due to high streaming velocities or low viscosities. There are various different types of turbulent flows that are distinguished by the properties of the flowing medium (water, air, plasmas, ...) or by the geometry and type of the boundaries (open, closed, moving, ...). However, turbulence has universal properties that motivates a systematic study of selected flow prototypes. Results might than be applied to similar systems.

In this course we are not only dealing with turbulence but with turbulent transport that is the motion of particles in a turbulent environment. It is not hard to imagine that this problem appears very often in environmental and industrial settings. Examples are the motion of pollutants in the atmosphere, the transport of organic material in the oceans, mixing devices, pumping of particle suspension through pipes or the motion of dust grains in protoplanetary discs.

The importance of turbulent transport for many open problems led to a intensive study of its statistical features. It has for example been found that turbulence leads to a

H. Homann (✉)
Laboratoire J.-L. Lagrange, Observatoire de la Côte d'Azur, CNRS,
Université Côte d'Azur, 06304 Nice, France
e-mail: holger@oca.eu

C. Marchioli (ed.), *Collective Dynamics of Particles*,
CISM International Centre for Mechanical Sciences 576,
DOI 10.1007/978-3-319-51226-6_2

fast separation of neighboring particles and hence to very efficient mixing. Another feature is that particles that have inertia decouple from the dynamics of the fluid thereby creating spatial inhomogeneities called clusters. Recent studies where related to turbulent fluctuations that can throw particles with high speeds against each other. This phenomenon called caustics and has to be considered for the problem particle collisions.

Most of the collected data and theory is on the dynamics of very small particles, so called *point-particles*. This is because in applications transported particles are indeed often much smaller than any scale of the fluid motion and because the smallness assumption simplifies significantly their study: if a particle is much smaller than any scales on which the flow varies and if in addition it is moving only slowly with respect to the fluid the particle will be surrounded by a Stokes flow (see Fig. 1 (left)). The corresponding flow structure originates only from viscous and pressure stresses and has been calculated analytically. The Stokes flow is symmetric and the inflow does (by definition) not vary on scales comparable to the diameter of the particle; the inflow is said to be uniform. Any inertial effect or inhomogeneity of the fluid is negligible in this limit and the drag on the particle is given by

$$\frac{d\boldsymbol{v}}{dt} = -\frac{1}{\tau_p}(\boldsymbol{v} - \boldsymbol{u}), \tag{1}$$

where $\boldsymbol{u}$ is the fluid velocity, $\boldsymbol{v}$ the particle velocity and $\tau_p = 2\,\rho_p\,a^2/(9\,\rho_f\,\nu)$ its response time (ρ_p is the particle density, ρ_f the fluid density, a the radius of the particle and ν the kinematic viscosity of the fluid). τ_p is the time that a particle needs to relax to the velocity of the fluid and is a measure of inertia of the particle. In the particle equation (1) the fluid velocity u means more precisely the velocity at the position of the particle. Taken literally, this definition has of course no sense (the fluid is sticking to the surface of the particle) but for a Stokes flow a particle is

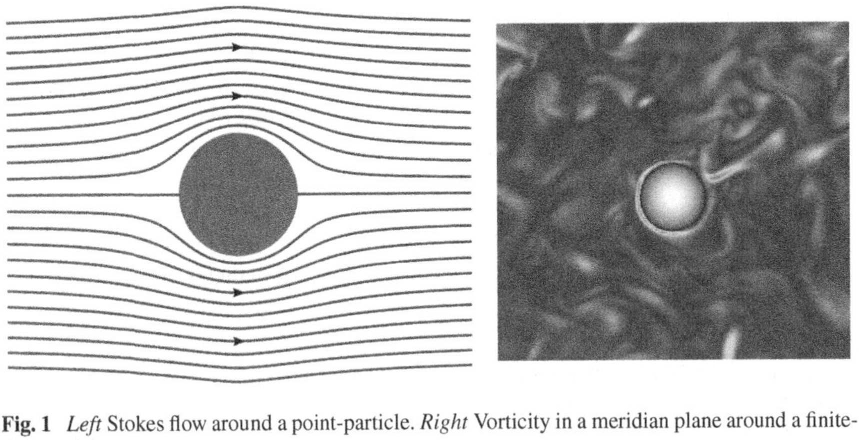

Fig. 1 *Left* Stokes flow around a point-particle. *Right* Vorticity in a meridian plane around a finite-size particle

facing a flow which is uniform at distances large compared to its diameter. In Fig. 1 (left) we could take any value for $\boldsymbol{u}$ at the boundary of the shown box. The nice thing about Eq. (1) is that it is just an ordinary differential equation for a given $\boldsymbol{u}$. This makes both analytic and numerical studies much simpler than studies of the transport of finite-size particles as we will see soon. Direct numerical simulations of turbulent point-particle suspensions can nowadays faithfully integrate the trajectories of billions of particles. Resulting data sets provide thus the basis for high precision statistical studies.

If one tries to model the force on bigger particles one encounters not only serious mathematical but also conceptual problems. The computation of the force in the point-particle approach relies (as we have seen) on the definition of a relative velocity of the particle with respect to the fluid. This very definition is not obvious if possible at all for a finite-size particle in turbulent flows. Particles that are larger than the Kolmogorov scale can be surrounded by a complicated flow structure (see Fig. 1 (right)). The fluid velocity in the vicinity of the particle can vary on scales that are smaller than the particle diameter. It is completely unclear which single velocity one should take from Fig. 1 (right) to define a meaningful velocity seen by the particle.

Such conceptual problems are one reason why the understanding of turbulence seeded with big particles is still in its infancy. Most of the works rely on the generation and exploitation of experimental or numerical data sets. But even the production of precise data is more complicated than for suspensions of very small particles. Numerically, one has to account for the interaction of the particle with the fluid which requires a precise modeling of the no-slip boundary condition at the surface of the particle and an adequate resolution of its boundary layer. Experimentally, one has to find ways to precisely track both the motion of the large and small particles. All this difficulties are responsible for an unpleasant situation: Even fundamental questions such as if turbulence is enhanced or attenuated by the presence of big particles is still under discussion. We will later come back to those open questions.

This course shall serve as an introduction to the topic of finite-size particle dynamics. Finite-size here means more precisely that either the diameter is of the order or larger than a characteristic scale of the flow (i.e. the Kolmogorov scale in turbulence) or that it has an important relative velocity with respect to the fluid so that it creates a wake. We can of course not discuss all possible flow types, open questions, numerical, experimental and analytic approaches. This is not the aim of this course so that we will only discuss selected problems and results. The main focus will be on the motion of finite-size particles in homogeneous isotropic turbulence. From time to time we will also consider the dynamics of sedimenting particles in a still fluid. This course includes a section that presents different approaches how the motion of finite-size particles and their fluid-particle interaction can be solved numerically. For experimental techniques the reader is asked to look at the course of Mickaël Bourgoin. The hope is that this introduction provides sufficiently many ideas, reflections and open questions that motivate the reader to get deeper into the story. A good starting point would be to read the cited publications as in each case only one or two of their results can be mentioned here.

This course is organized as follows: We will start with a very brief introduction to turbulence. Then, some selected numerical methods are discussed that are used for direct numerical simulations of the transport of finite-size particles. After that, an important place is reserved for the study of differences in the dynamics of small and large particle. Hereby, the modification of the carrier flow by finite-size particles will be presented. The last section deals with collective effects of many particles that explicitly arise from a large particle diameter.

2 Basics of Turbulence

In this section we will introduce some basic properties of turbulent flows that are important for the subsequent sections. We will focus on incompressible flows obeying to the Navier–Stokes equations

$$\partial_t \boldsymbol{u} = -\boldsymbol{u} \cdot \nabla \boldsymbol{u} - \nabla p + \nu \nabla^2 \boldsymbol{u} + \boldsymbol{f}_e, \quad \nabla \cdot \boldsymbol{u} = 0, \tag{2}$$

$\boldsymbol{u}$ denoting the velocity field, p the pressure, ν the kinematic viscosity, and $\boldsymbol{f}_e$ a possible external force maintaining the flow. The degree of turbulence can be characterized by the Reynolds number $Re = u_{rms} L/\nu$, L being the forcing scale and u_{rms} the root-mean-square value of the velocity. Often also the Taylor Reynolds number $R_\lambda = \sqrt{15\, Re}$ is used.

One of the properties of high Reynolds number turbulence is the creation and interaction of an enormous range of spatial and temporal scales. Let us take the example of a cloud and assume that it has a total size of one hundred meters. Its turbulent motion is driven at comparable scales by convection and shear. In three dimensional turbulence the energy contained in these large scales cascades to smaller and smaller scales. At very small scales of the order of one millimeter the viscosity of air transforms the kinetic energy into heat. In our cloud example we have thus a scale separation from 100 m to 1 mm, thus 10^5 different interacting scales involved. The largest scale is called *integral scale* L and the smallest scales is called *Kolmogorov scale* η. The mentioned energy cascade from large to small scales is a basic property of 3D turbulence. In between the two limiting scales (100 m and 1 mm) a purely inertial transfer of kinetic energy is taking place. In this interval of scales, called *inertial range of scales*, neither the particular mechanism of driving at large scales nor the dissipation mechanism at small scales is important so that the flow is more or less homogeneous, isotropic and approximately scale invariant.

Many scientific activities are devoted to the study of the idealized flow type of homogeneous isotropic turbulence (HIT). In this case statistical quantities such as velocity gradients are invariant under translation and rotations. Practically, one can think for example of a small parcel of a cloud in a co-moving (with the average velocity) frame of reference. In order to get a rough idea of the flow structure of HIT a snapshot of vorticity is shown in Fig. 2. This picture is taken from a direct numerical simulation (see subsequent section) of three-dimensional incompressible

Fig. 2 Rendering of vorticity in a small slice of a direct numerical simulation of homogeneous isotropic Navier–Stokes turbulence with 4096^3 grid points ($R_\lambda = 730$). The *middle figure* applies a zoom of 4 while the *right figure* applies a zoom of 16 compared to the *left figure*

turbulence with periodic boundary conditions. One can see that HIT is composed of regions of strong vorticity (and strong dissipation in its vicinity). Looking in more detail into these regions (Fig. 2 (middle and right)) reveals that they consist of entangled vorticity filaments obviously interacting with each other. The diameter of such filaments is of the order of 10η and they belong to the smallest structures of HIT. These filaments are also present in other types of turbulent flows.

The fact that there is no characteristic scale within the inertial range of scales implies that scale (l) dependent quantities such as velocity differences along a separation $\boldsymbol{l}$

$$\delta_l u = (\boldsymbol{u}(\boldsymbol{x}+\boldsymbol{l}) - \boldsymbol{u}(\boldsymbol{x})) \cdot \boldsymbol{l}/|\boldsymbol{l}| \tag{3}$$

behave as power laws. Indeed, the (second-order) structure function $S_2(l) = \langle(\delta_l u)^2\rangle$ scales as $S_2(l) \sim l^{\zeta_2}$ within the inertial range and the extend of this spatial scaling range can in turn be estimated from S_2. Two examples of S_2 are plotted in Fig. 3

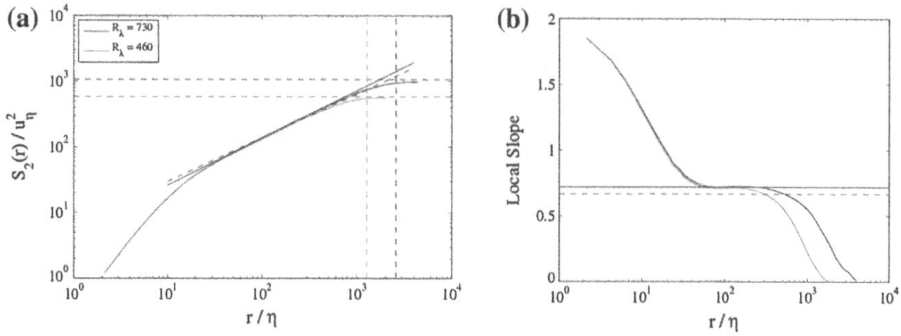

Fig. 3 *Left* Second order structure function S_2 in HIT for two different Reynolds numbers $R_\lambda = 730$ ($Re = 35000$) and $R_\lambda = 460$ ($Re = 14000$). *Right* Local slope of S_2 that is its logarithmic derivative $d\log(S_2)/d\log(r)$. The inertial range of scales can be defined as the interval where the local slope of S_2 is zero. Its scaling exponent ζ_2 corresponds to the value of this plateau (Bitane et al. 2013)

(left) for the simulation shown in Fig. 2 and for a simulation with a smaller Reynolds number. The main message is that at sufficiently high Reynolds number an inertial range of scales clearly exists but that Re has to be sufficiently high ($Re \approx 10000$, corresponding to a grid resolution of 2048^3 grid points).

Another important fact is that the scaling exponent ζ_2 is much smaller in the inertial range (see Fig. 3 (right)) than in the dissipation range where the flow is smooth ($\zeta_2 = 2$). The size of the vorticity filaments can roughly be seen as a landmark between the small and inertial range scales. Turbulence is thus not smooth withing the inertial range of scales. However, these are the scales we are talking about if particles are meant to be of finite-size!

In this course we will consider the transport of particles having a size equal to or larger than the vorticity filaments. The basic equations of motion of a particle of any size read

$$m_p \ddot{\boldsymbol{X}} = \int_S \nabla \cdot \boldsymbol{T} dV = \int_{\partial S} \boldsymbol{T} dS, \tag{4}$$

$$\frac{2}{5} m_p a^2 \dot{\boldsymbol{\Omega}} = \int_{\partial S} \boldsymbol{n} \times (\boldsymbol{T} \cdot dS), \tag{5}$$

$\boldsymbol{T}$ being the stress tensor

$$\boldsymbol{T} = -p\boldsymbol{I}_3 + \nu(\nabla \boldsymbol{u} + \nabla \boldsymbol{u}^T). \tag{6}$$

The first equation (4) determines its translational motion and the second equation (5) its rotation. The particle Reynolds number $Re_p = D_p V_p/\nu$ usually involves the particle diameter D_p, its relative velocity V_p and the kinematic viscosity. The velocity $\boldsymbol{u}$ appearing in (4) is a solution to the Navier–Stokes equations (2) with a no-slip boundary condition at the particle surface. This is a coupled non-linear system of partial differential equation that shows the complexity of the problem.

Let us conclude this section by stressing that homogeneous isotropic turbulence is one type of turbulent flow on which physicists often concentrate. However, one should not forget that in real flows often a multitude of physical processes is contributing to its evolution. Concerning a cloud, for example radiation, latent heat and electricity play for example a role. Also there are of course other archetypal and important flows such as convecting, rotating, pipe-, wall- and channel flows.

3 Numerical Methods for Finite-Size Particles

Finite-size particles in turbulence are surrounded by a complicated flow structure (recall Fig. 1 (right)) for which no general analytic solution can be found. Indeed, finding a simple equation modeling the force on such a particle in the spirit of (1) is part of ongoing research activities. In order to get insight into the dynamics of finite-size particles one can perform direct numerical simulations (DNS). DNS means that the basic equations (2), (4), and (5) are solved directly at all scales and no modeling or simplifications are used. This is especially adequate for systems that are not well understood. In the case of large particle dynamics this kind of simulations has become a valuable tool because the computing power of supercomputers and the performance of algorithm increased to such an extend that DNS of suspension of big particles are nowadays possible.

Several different strategies have recently been proposed in order to perform DNS of spherical particles. Most of them make use of a homogeneous grid that covers the complete physical domain and resolves all turbulent scales. The principle idea is to use a solver of the Navier–Stokes equations and to impose in some way the no-slip boundary condition at the surface of the particle. Let us now mention some of those strategies:

"Physalis", developed by Prosperetti and Oguz (2001) and extended by Naso and Prosperetti (2010), uses the analytic solutions to the Stokes equation for the viscous boundary layer in the vicinity of the particle. This Stokes solution is used to prescribe the velocity at the nearest surface points of the grid. Iteratively, the analytic stokes solution and the outer solution on the homogeneous grid are matched. That way the no-slip condition is imposed in an exact manner if the grid spacing is much smaller than the viscous boundary layer.

The "Overset Grid" technique proposed by Burton and Eaton (2005b) consists in using and matching two different grids. A spherical one encloses the particle surface and is designed to resolve the boundary layer very accurately and a second homogeneous grid that covers the entire domain. A third-order interpolation is used to transfer values from one grid to the other in order to couple the two solutions.

The "force coupling" method of Maxey and Patel (2001) solves the Navier–Stokes equations including a body force that originates from the transported particles. This body force is computed from a low-order multipole expansion of the force exerted by a particle on the fluid. From this expansion they retain the monopole term including buoyancy and forces from the displaced fluid and the dipole term including straining and rotation effects. Due to numerical manageability they replace the Dirac delta function by a Gaussian envelop of a width which is connected to the size of the particle. The no-slip boundary condition is therefore only approximately fulfilled.

The "immersed boundary" approach solves the Navier–Stokes equations in the entire domain (usually on a uniform grid) including the particles. The no-slip boundary condition is imposed via additional forces acting on the fluid in the volume occupied by the particles. This technique has been used together with finite-difference schemes (Uhlmann 2005) and pseudo-spectral schemes (Homann et al. 2013). Let us now look at the latter in more detail.

The main idea consists in combining a Fourier pseudo-spectral method for the fluid with an immersed-boundary technique to impose the no-slip boundary condition on the surface of the particle. Pseudo-spectral means that spatial derivatives arising in (2) are computed in Fourier-space while convolutions from the non-linear term are computed in real space. A Fast Fourier Transformation (FFT) is used to switch between the two spaces. Spectral schemes are a standard tool of DNS of HIT because of their accuracy.

The no-slip boundary condition at the particle surface is treated via an immersed boundary technique. For a particle at rest this condition reads

$$\boldsymbol{u}(x,t) = 0 \quad \text{for } \boldsymbol{x} \in \partial\Omega_{\mathrm{p}}, \tag{7}$$

for a particle at rest and center $\boldsymbol{x}_0$. Here, $\Omega_{\mathrm{p}} = \{\boldsymbol{x} : |\boldsymbol{x} - \boldsymbol{x}_0| \leq D_p/2\}$ denotes the volume occupied by the particle. The immersed boundary technique consists in introducing in the right-hand side of the Navier–Stokes equation (2) a force $\boldsymbol{f}_b(\boldsymbol{x},t)$ associated to the constraint defined by the boundary condition (7). The full problem (2)–(7) can then be rewritten as

$$\partial_t \boldsymbol{u} = L(\boldsymbol{u}) + \boldsymbol{f}_b, \quad \nabla \cdot \boldsymbol{u} = 0, \quad \text{for } \boldsymbol{x} \in \Omega, \tag{8}$$

where $L(\boldsymbol{u})$ denotes the right hand side of (2). There are now different ways to compute the force $\boldsymbol{f}_b$. One of them is called direct forcing method (see Fadlun et al. 2000) and directly imposes the particle velocity to the grid points in the volume occupied by the particle. As a uniform grid does not coincide with the particle surface some sort of interpolation of the surface velocity to the nearest grid points has to be used. This results in an effective smoothing of the particle surface making its surface slightly porous.

Here might be the right place to stress today's limits of direct numerical simulations. Modern efficient numerical codes which include optimized algorithms and parallelization run on thousands of computing cores. They use the fastest supercomputers in the world and consume millions of CPU hours per simulation. Nevertheless, the accessible parameter space is still limited. The Reynolds number of the flow is usually of the order of one thousand and also the number of particles does not exceed a few thousand. This is one reason why the amount of available data on this topic is also still limited. One should mention that experiments have the advantage to reach higher Reynolds and particle numbers but often suffer from other difficulties.

4 Finite-Size Effects of Individual Particles

In the introduction we saw that the flow structure around finite-size particles in turbulent flows is in general different from that of point particles. However, we do not know yet if their dynamics differ. But in fact they do. In this section we are going to present finite-size features of particle transport that definitely distinguish large from small particle dynamics.

4.1 Slip Velocity

Let us start by discussing the slip velocity of large particles. Slip velocity means the relative velocity of a particle with respect to the fluid. In the introduction we saw that this velocity difference had a simple meaning for point particles. They are so small that the surrounding flow is nearly uniform at large distances from the particle. In this region, every fluid parcel is moving with the same velocity relative to the particle. This velocity is obviously the slip velocity.

A finite-size particle in turbulence is (by definition) facing a flow that varies on scales smaller than the particle diameter (recall Fig. 1 (right)). There is no such region of uniform fluid velocity. However, also a big particle is somehow moving through the fluid and creating eventually even a clearly visible wake. There might be therefore at least some hope that a meaningful definition of the slip velocity exists. Because of the turbulent fluctuations spatial averages might be a good candidate. Lucci et al. (2010) and Kidanemariam et al. (2013) define the fluid velocity seen by the particle to be the average velocity on a concentric shell around the particle. They take a shell with a radius of the order of the particle diameter. The idea is that at this distance the flow is only weakly influenced by the boundary layer of the particle. The slip velocity is then the difference of this averaged velocity and the particle velocity.

Cisse et al. (2013) computed the shell averaged velocity as a function of the shell radius. They introduced a two dimensional coordinate system consisting of the slip direction and a normal direction. The slip direction $\vec{e}_r(t)$ is computed from the fluid flux through each shell as

$$\vec{e}_r(t) = \vec{\Phi}_r(t)/|\vec{\Phi}_r(t)|, \quad \text{where } \vec{\Phi}_r(t) = \int_{S_r} \left(\vec{u}(\vec{x}, t) - \vec{V}_p(t)\right) \cdot \vec{n} \, d\vec{S}, \tag{9}$$

where $\vec{u}$ and $\vec{V}_p$ are the fluid and the particle translational velocity, respectively, $\vec{n}$ is the unit vector normal to the shell. In other words, on each shell an average of the direction weighted by the fluid mass flux is performed, so that $\vec{e}_r$ points in the direction of the flux on the shell at distance r. This choice is physically motivated as the fluid enters such a shell upstream and exits it in the wake. For a Stokes flow, $\vec{e}_r$ would be independent of r and exactly parallel to the slip direction.

Having the direction $\vec{e}_r$ defined, one can project the velocity difference $\vec{u} - \vec{V}_p$ onto it and perform a time average to construct the mean velocity profile of the flow relative to the particle

$$U_{rel}(\rho, z) = \left\langle \left(\vec{u}(\vec{x}, t) - \vec{V}_p(t)\right) \cdot \vec{e}_r \right\rangle, \tag{10}$$

with $z = (\vec{x} - \vec{X}_p(t)) \cdot \vec{e}_r$ and $\rho = [|\vec{x} - \vec{X}_p(t)|^2 - z^2]^{1/2}$. The coordinates z and ρ, which are defined at each instant of time, are in the direction of $\vec{e}_r$ and perpendicular to it, respectively. By rotational symmetry around the axis defined by $\vec{e}_r$, the mean profile U_{rel} depends on z and ρ only and not on the angle.

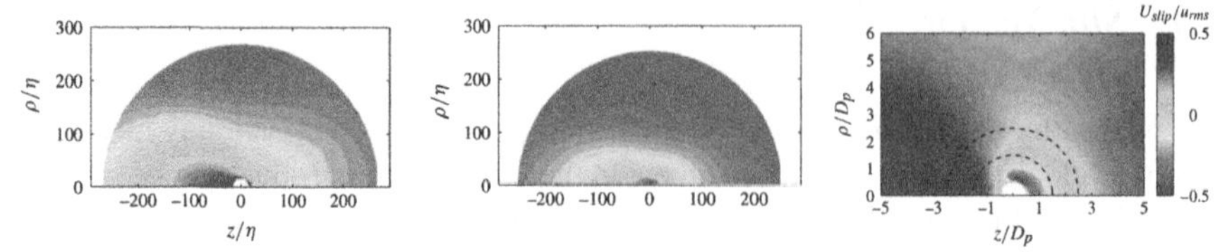

Fig. 4 *Left* Temporal- and angle-averaged fluid relative velocity projected on the direction of motion for $D_p = 34\eta$ (*left*) and for a tracer (*middle*). *Right* slip velocity U_{slip} for $D_p = 34\eta$ defined as the difference between the mean relative velocity profile U_{rel} around the particle and that around a tracer; the two *black dashed circles* represent distance equal to D_p and $2D_p$ from the particle surface (Cisse et al. 2013)

The relative velocity profiles (see Fig. 4 (left)) have a marked asymmetry resembling the inflow and wake structure of a particle facing a uniform flow. However, performing the same computations for a tracer particle moving in a similar flow produces a similar velocity profile (see Fig. 4 (middle)). The problem is that a tracer has by definition no slip velocity or wake. Indeed, the measured velocity profile is a consequence of the intrinsic asymmetry of the longitudinal velocity increments (3) in HIT and the specific choice of the coordinate system. In HIT, velocity differences in the direction of the separation increase with the separation and are skewed. Cisse et al. (2013) therefore proposed to study the difference of the mean velocity profiles of a finite size particle and that of a tracer (see Fig. 4 (right)). This difference shows strong variations close to the particle (up to twice its diameter) while it is approximately constant at further distances. This constant is then defined to be the average slip velocity.

A completely different strategy to define a relative velocity is pursued by Bellani and Variano (2012). They define a *stochastic* slip velocity using the variance of the fluid and particle velocity

$$U_{rel} = \left(\langle \boldsymbol{u}'^2 \rangle - \langle \boldsymbol{v}'^2 \rangle\right)^{1/2} . \tag{11}$$

Here, the primes denote the fluctuations with respect to the mean $\boldsymbol{u}' = \boldsymbol{u} - \langle \boldsymbol{u} \rangle$. An advantage of this definition is its simplicity and direct applicability to experiments. However, from its very definition it is clear that this only gives an average and no local definition.

We see from this that already the definition of slip velocity is a subtle problem for finite-size particles. Next we will look at the modification of turbulence in the vicinity of an individual particle.

4.2 *Modification of Turbulence*

It is well known that finite-size particles in a uniform flow create a boundary layer and wake. One may ask how this modifies the properties of the turbulent flow. How do for example particles change the energy dissipation rate in its vicinity?

Burton and Eaton (2005a) performed 3D DNS of freely decaying HIT in which they put one fixed particle with a diameter comparable to the Kolmogorov length scale. By means of ensemble averages they find that the local energy dissipation is increased close to the particle compared to the outer turbulence. The particle displaces the fluid, creates a boundary layer and increases thereby velocity gradients. Further away, there is some indication that at intermediate distances the dissipation rate might be reduced. This has also been found by Naso and Prosperetti (2010) who studied the same situation as Burton and Eaton but in stationary (forced) HIT (see Fig. 5 (left)). A reduced energy dissipation rate means that the particle attenuates turbulent velocity fluctuations at small scales (the energy dissipation is a small scale quantity). The particle influences obviously scales smaller than its diameter.

Cisse et al. (2013) studied the same question but for a moving particle. They performed numerical measurements around a neutrally buoyant (having the same density as the fluid) sphere. Using their definition of the slip direction (see previous section) they distinguished the energy dissipation rate in the upstream and downstream flow (see Fig. 5 (right)). In the upstream flow the dissipation is strongly increased. Downstream, at distances from $0.2D_p$ to $1D_p$, the dissipation is reduced compared to the bulk turbulence. It is thus the wake of the particle that attenuates turbulence. However, very close to the particle (in the viscosity dominated boundary layer) the dissipation is even downstream increased.

These results concern particles moving with moderate slip velocity through the fluid. Rough estimates of their particle Reynolds numbers range from 20 to 200. At these speeds the wake of the particle is still laminar. But higher speeds (for bigger

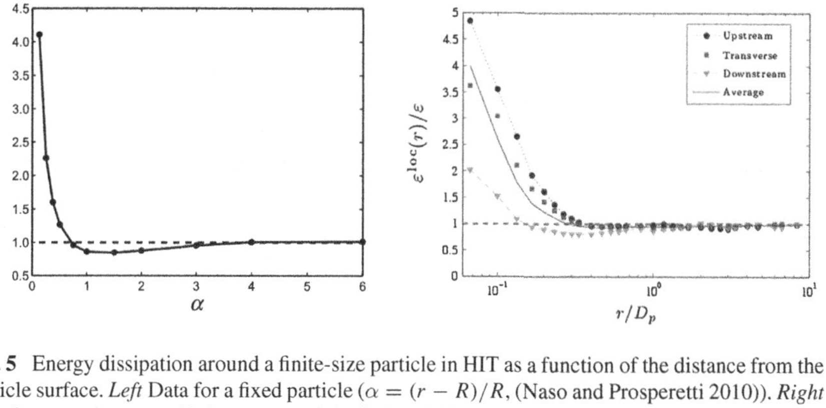

Fig. 5 Energy dissipation around a finite-size particle in HIT as a function of the distance from the particle surface. *Left* Data for a fixed particle ($\alpha = (r - R)/R$, (Naso and Prosperetti 2010)). *Right* Data for a moving neutrally buoyant particle distinguishing contributions from up- and downstream (Cisse et al. 2013)

or heavier particles) would imply the creation of turbulent fluctuations in the wake and those might lead to an increase of the energy dissipation rate.

Let us now turn to the drag force on a large particle facing turbulent fluctuations. It is clear that this quantity is crucial for the possible modeling of large particle dynamics via a simple equation such as (1).

4.3 Drag Force in Turbulent Flows

The drag force on a particle moving at a constant speed U_c (see Fig. 6) through a fluid at rest has been studied for a long time. The corresponding drag coefficient $C_D = 8F/(\rho U_c^2 \pi D_p^2)$ is a non-trivial function of the particle Reynolds number Re_p (see Fig. 7 (left)). Analytic results have only been obtained for small Re_p. Other formulas such as the one of Schiller and Naumann (1933)

$$C_D(Re_p) = \frac{24}{Re_p}(1 + 0.15\, Re_p^{0.687}) \tag{12}$$

are only reasonable fitting functions to experimental and numerical data up to a certain Re_p.

At small Re_p the drag is increasing linearly (C_D is decreasing) according to the Stokes drag. From $Re_p \approx 10^3$ to $Re_p \approx 10^5$ it increases quadratically so that C_D becomes flat. Associated to these different drag regimes are specific flow patterns around the particle. For small Re_p one finds a steady, approximately symmetric flow resembling the Stokes flow. At intermediate Re_p the flow is still steady but a recirculation region appears in the wake. In the quadratic drag regime the wake is turbulent (see also Fig. 6). At a even higher Re_p the boundary layer is said to become turbulent (a regime that we will not consider).

We see that the drag force and the according flow patterns are already complicated in the case of an uniform inflow. The question arises what happens if the inflow is turbulent (see Fig. 8). Clearly, turbulent velocity fluctuations will interact with the boundary layer and wake of the particle. The question is whether one can also apply the empirical formula (12) to the case of a turbulent flow interacting with a particle or will the drag force be modified?

Bagchi and Balachandar (2003) analyzed this question by means of different numerical simulations. They either placed a finite-size particle or point particle into

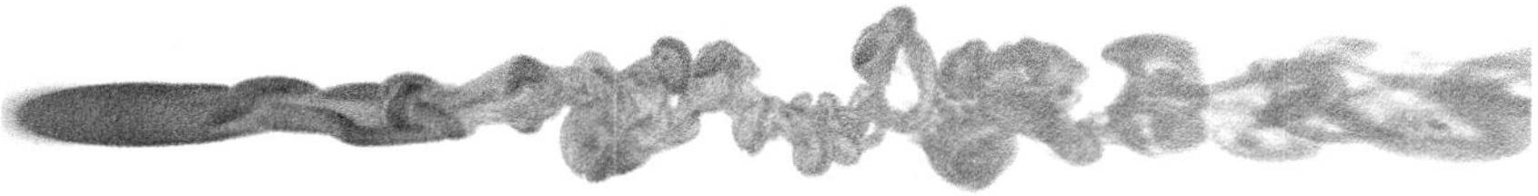

Fig. 6 Volume rendering of vorticity of a uniform flow passing a spherical particle with $Re_p = 1000$

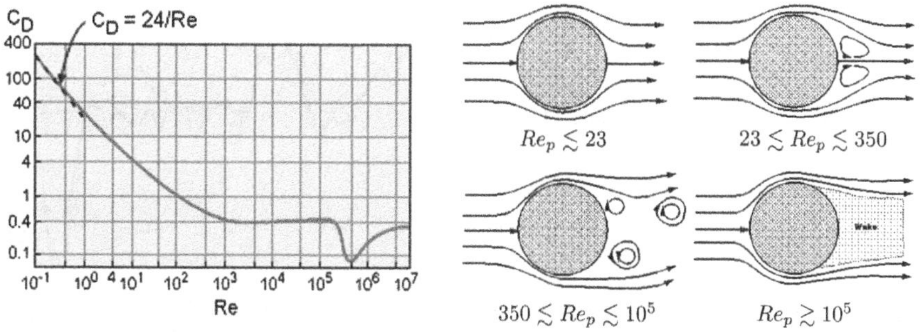

Fig. 7 *Left* Drag coefficient of a spherical particle facing a uniform inflow. C_D. *Right* Flow structure depending on the particle Reynolds number

Fig. 8 Volume rendering of vorticity of a turbulent flow with average speed U_c passing a spherical particle with $Re_p = 400$ and $I = u_{rms}/U_c = 0.12$

the same turbulent inflow. They found that already for a particle size of $d/\eta = 1.5$ only the large scale variations are correctly reproduced while the small scales are clearly not captured (see top panel of Fig. 9). This can be understood by considering the large scale variations as a slowly varying inflow. If the inflow velocity changes on time-scales much larger than the time it takes the fluid to passe the particle, the latter faces a quasi-stationary headwind. The small scale fluctuations are more complicated to understand and model. They are not include in standard drag correlations for uniform inflows.

The agreement of the finite-size DNS and the point-particle simulations is getting even worse when increasing the particle size (see middle and bottom panel of Fig. 9). Bagchi and Balachandar (2003) checked whether this can be improved by taking additional force terms into account. Indeed, Maxey and Riley (1983) and Gatignol (1983) derived the equation

$$\frac{d\boldsymbol{v}}{dt} = -\frac{1}{\tau_p}(\boldsymbol{v} - \boldsymbol{u}) + \beta\frac{D\boldsymbol{u}}{Dt} - \sqrt{\frac{3\beta}{\pi\tau_p}}\frac{d}{dt}\int_0^t \frac{\boldsymbol{v} - \boldsymbol{u}}{t - \tau}d\tau \tag{13}$$

with $\beta = 3/(2\rho_p/\rho_f + 1)$ (ρ_p and ρ_f being the mass density of the particle and fluid, respectively). The second and third term on the right hand side are called added mass term and history term, respectively. However, Bagchi and Balachandar (2003) showed that the inclusion of the added mass and even the history term did not improve the agreement of the point model with the finite-size simulations. This is a bit surprising, as Daitche (2015) analyzed in detail the role of the history force

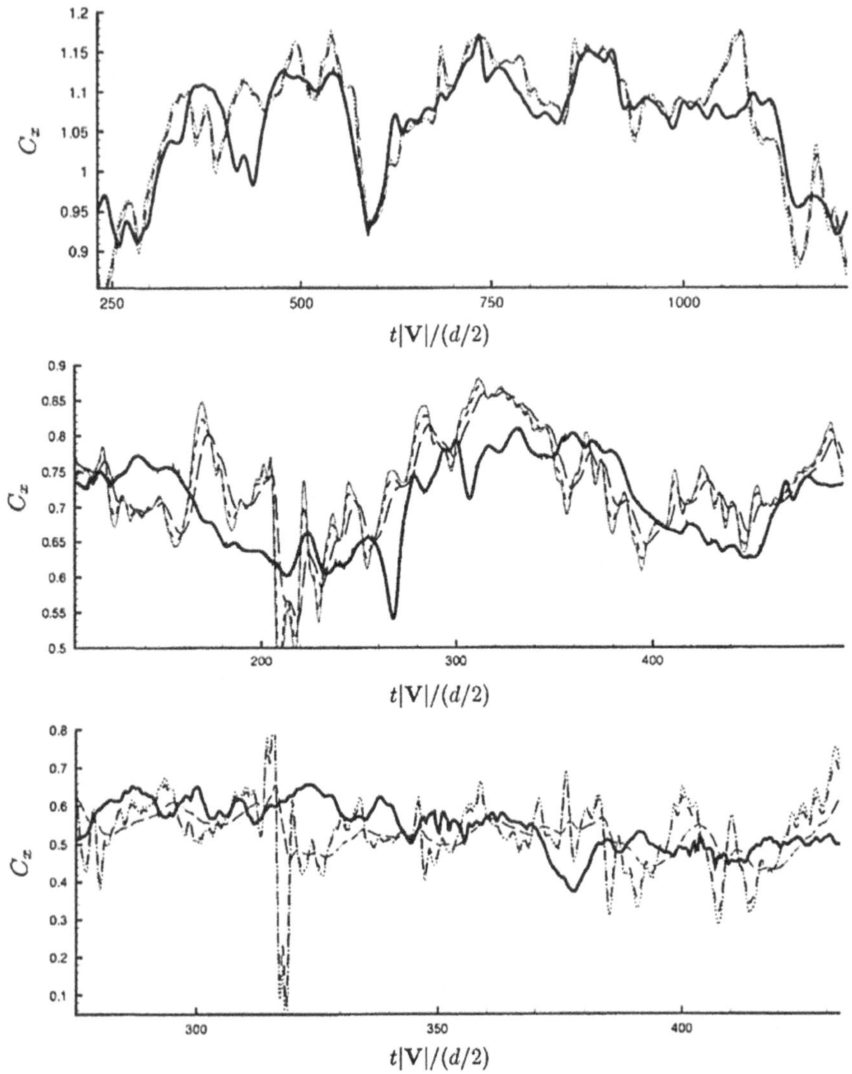

Fig. 9 Time series of the stream-wise force component of a spherical particle facing a turbulent inflow. *Solid line* DNS of a finite-size particle (*top* $d/\eta = 1.5$, $Re_p = 107$; *middle* $d/\eta = 3.8$, $Re_p = 261$; *bottom* $d/\eta = 9.6$, $Re_p = 609$) facing a turbulent inflow of $I = 0.1$. *Long dashed line* Schiller and Naumann law (12). *Short dashed line* plus the inertial force. *Dotted line* plus the history force (Bagchi and Balachandar 2003)

in a turbulent flow and found that its relative contribution compared to the Stokes drag scales as a/η. One could have thus expected the importance of this force term to grow with the particle size.

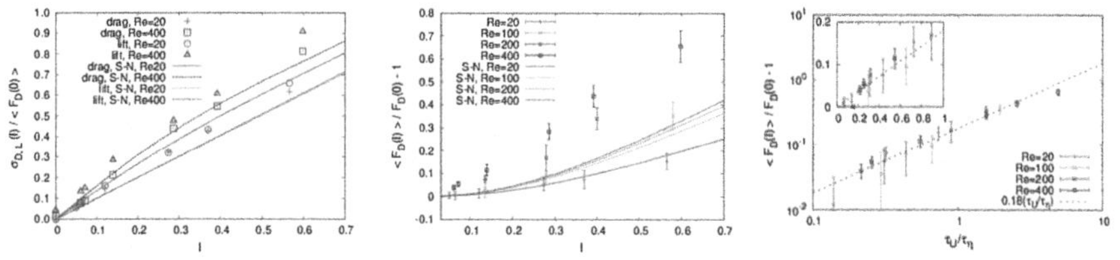

Fig. 10 *Left* Standard deviation of the drag force in free-stream turbulence for different turbulent intensities I. The *lines* correspond to a simple model assuming a quasi-static inflow speed. *Middle* Mean drag force. *Right* Mean drag force this time as a function of τ_U/τ_η = particle passing time/turbulent dissipation time scale (Homann et al. 2013)

Kim and Balachandar (2012) found indications that the drag force might be increased in turbulent compared to uniform flow. And here, increased means beyond that what one would expect from quasi-static changes of the inflow speed. Indeed, a slowly varying inflow speed already leads to an increase of the average drag force due to its non-linear dependence on the inflow speed (see (12)). The observed additional increase, however, was within the error bars.

Let us note here that this fluid dynamical problem has three independent parameters: the fluid Reynolds number $Re = u_{rms}L/\nu$, the particle Reynolds number $Re_p = U_c D_p/\nu$ and the turbulent intensity $I = u_{rms}/U_c$. Other non-dimensional quantities such as D_p/η are consequences of them.

Later, Homann et al. (2013) redid DNS of a similar situation (fixed sphere in turbulent headwind) and found that the standard deviation of the fluctuations of the drag force follow approximately the prediction of quasi-static changes of the inflow speed (see Fig. 10 (left)). The large scale fluctuations of the inflow translate into a shaking of the particle that is captured by the quasi-static assumption. However, the mean drag force was significantly higher than what one would expect (see Fig. 10 (middle)). From the right panel in Fig. 10 Homann et al. (2013) claimed that this increase might be due to small scale fluctuations of the inflow: Once the drag force is plotted as a function of the ratio of two small-scale time-scales all data collapse to one single straight line. The two time-scales are the time it takes the flow to pass the particle and the turbulent dissipation time scale. The mean drag force might be increased by small scale fluctuations that modify the boundary layer in such a way to increase the velocity gradients. Steeper gradients in turn lead to larger forces.

There are of course other situations than turbulent flows in which things change for finite-size particles. One of them is the motion of a particle sedimenting under the action of gravity in a still fluid. Let us look at this now. This will serve as a preparation for the complicated case of the interaction on many sedimenting finite-size particles discussed later.

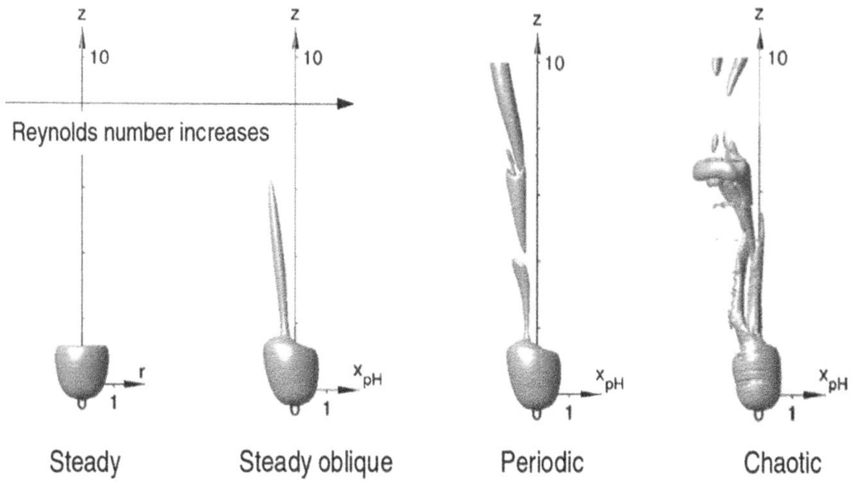

Fig. 11 Flow structure (vorticity) of a spherical particle sedimenting in a quiescent fluid. The particle Reynolds number increases from left to right. Remark that the wake structure is not vertical (Uhlmann and Dušek 2014)

4.4 Sedimenting Particles

We will consider the case of a single finite-size particle sedimenting in quiescent fluid. Despite the fact that the wake structure depends on the particle Reynolds number one finds that depending on Re_p particles are not just falling along vertical lines. Indeed, for intermediate Re_p particles move on straight but oblique trajectories (see Fig. 11). Even if the wake is steady it can give rise to a horizontal velocity. One would of course not expect this feature for point-particle as they are (by definition) moving in a symmetric (Stokes) flow.

5 Collective Effects of Finite-Size Particles

This section is devoted to collective effects in the particle dynamics, i.e. effects arising from the interaction of many (at least two) particles. Let us already distinguish two different classes of collective effects: Those simply arising from a superposition of the imprints of individual particles onto the flow in very diluted systems and those originating from the hydrodynamical interactions of particles.

Collective effects are worth studying because the admixture of impurities to a turbulent flow can have a significant impact onto the flow. It is for example well known that even a very small polymer concentration can lead to a significant drag reduction in oil pipes. A feature used in the petroleum business. Oil with polymers is a visco-elastic fluid for which models exists that allow for an exchange of the kinetic energy of the fluid and the elastic energy of the polymers. Polymers suppress large velocity gradients and lead in this way to a drag reduction.

Finite-size particles have a more complicated imprint on the flow. In turbulent settings its radius of influence is of the order of its diameter. This is true for neutrally buoyant particles while other particles with higher Reynolds numbers might produce wakes that are visible at even longer distances. It is not surprising that in suspensions of many particles these flow modifications can have a macroscopic effect on the flow structure. In very dilute suspensions (the simplest case) the individual particle contributions might simply add up so that modifications can be understood from the one-particle dynamics. But in denser suspensions particles and their respective wakes interact with each other thereby creating new effects. The aim of this section is to present a selection of collective effects of finite-size particles. Data and results on this topic are still quite rare as already the understanding of the dynamics of a single particle is lacking. So let us go step by step. Before considering a large number of particles we will study what happens if two particles come close to each other and interact hydrodynamically.

If two particles approach each other they squeeze out the fluid between them. This creates a repulsive force that is called lubrication force. Particles with smooth surfaces cannot touch a priori because this force goes to infinity at contact. Realty, however, might be different. The surface might be rough, van der Waals forces might act or the fluid assumption (inter-particle distance much larger than the mean free path) might break down. The lubrication force is also at play if close particles separate. Now, fluid has to be squeezed in so that particles slow down by transferring their kinetic energy to the fluid.

In the next paragraph we will see how the drag and lift forces on a particle are affected by the presence of a second particle in its vicinity.

5.1 Two Interacting Particles

The main question of this section is what happens to the drag force on a fixed particle in a uniform flow if a second particle comes close to it? There are two very different possible arrangements of the particle positions. They can be on a line in stream-wise direction or on a line perpendicular to it. In the first case we expect that the drag on the following particle is reduced in the slipstream of the leading particle (an experience made by cyclists and motorists). Let us now consider this configuration in more detail.

Zhu et al. (1994) experimentally measured the drag force on two particles that are aligned in stream-wise direction (see Fig. 12 (left)). The particles build a tandem with a fixed separation. They varied their distance in a uniform flow and measured the associated drag forces. They found that the trailing particle experiences a significantly reduced drag force that can be as small as one fifth of the drag force of an isolated particle (see Fig. 12 (right)). The closer the particles are the smaller is the drag. The reason for this is the wake of the leading particle in which the difference between

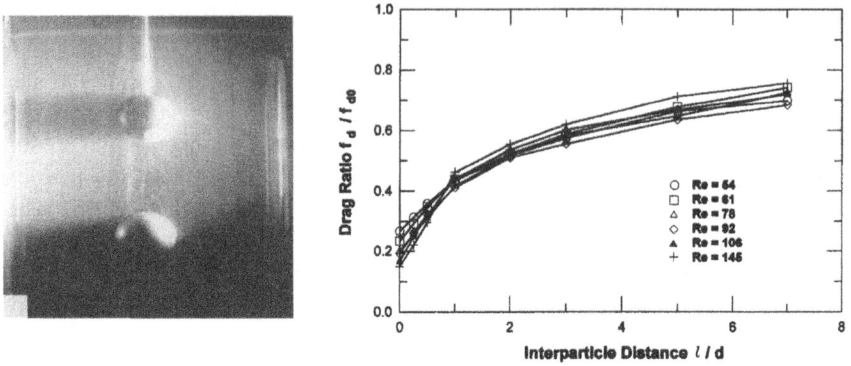

Fig. 12 *Left* Two particles aligned in stream-wise direction. *Right* Drag force on the trailing particle (Zhu et al. 1994)

the particle velocity and that of the fluid is reduced. This so called velocity deficit is zero at the particle surface and recovering only asymptotically the inflow value. The leading particle is entraining the fluid. The trailing particle will thus face a reduced headwind speed that in turn reduces its drag. This is quite obvious. However, as can also be seen in the right panel of Fig. 12, there is some non-trivial Reynolds number dependence of this drag reduction. At small distances from the particle the drag reduction is the larger the smaller is the particle Reynolds number. This tendency is reversed at larger distances. The cross-over happens at approximately at a distance of one diameter.

Before continuing let us stress the importance of this effect of drag reduction, called *wake attraction effect*. It is clear that this effect has the potential to bring particles together so that for example sedimenting particles will tend to accumulate. And it is also clear that a bunch of particles will have other dynamics than an individual particle. We will later come back to this point.

The work of Zhu et al. (1994) documents another interesting observation: Also the leading particle experiences a drag reduction when a trailing particle catches up. This reduction is quite small and reaches only ten percent at a distance of one diameter before it is slightly reincreasing. For the leading particle it is thus beneficial to have a particle following closely. The reason is that a part of the drag on a particle originates from its wake generation. The trailing particle obviously modifies the boundary layer and wake of the leading particle in such a way to reduce stresses.

Now let us come to the second configuration: We will put the particles side by side instead of one after another. Kim et al. (2006) fixed two of them as sketched in Fig. 13 (left) in a uniform flow and varied their Reynolds number and distance. They report two important findings: First, their individual drag force increases significantly for distances smaller than one diameter. The increase is larger the closer they are. The second particle has a sort of blocking effect on the fluid passing the first. The second important finding is that the lift force i.e. the force between the particles varies non-trivially as a function of the inter-particle distance. Close particles (closer than 1.5

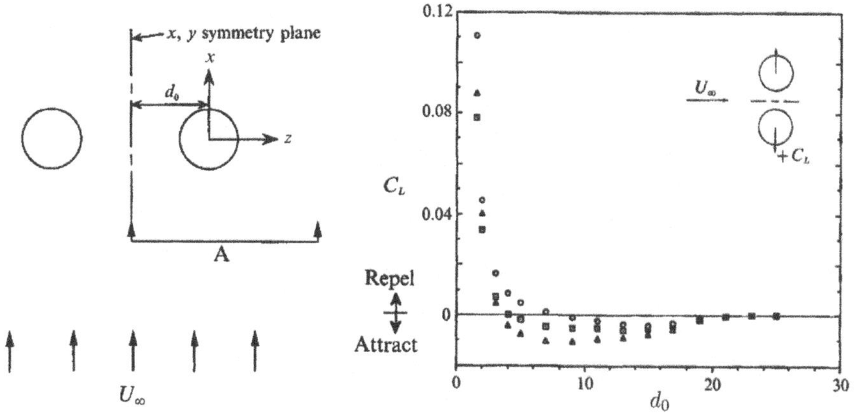

Fig. 13 *Left* Two particles aligned side by side. *Right* Lift force on the particles (Kim et al. 2006)

diameter) are repelled. The flow between them pushes them strongly apart. However, particles at distances of two to twenty diameter are slightly pulled together (see Fig. 13 (right)). This attraction effect increases with the particle Reynolds number. Roughly speaking this sounds like the Venturi effect where the static pressure decreases when the dynamics pressure increases. However, it is not obvious why this effect inverses at particle separations of the order of one diameter.

So far the two particles were fixed. Let us now see what happens to two sedimenting particles in still fluid (see Fig. 14). If they are vertically aligned the wake attraction effect will accelerate the trailing particle to catch up the leading one. They will collide which gives horizontal momentum to the particles and make them separate again. Their boundary layers and wake interact in a complicated manner leading to a tumbling motion. This dynamics is called *drafting, kissing, tumbling*.

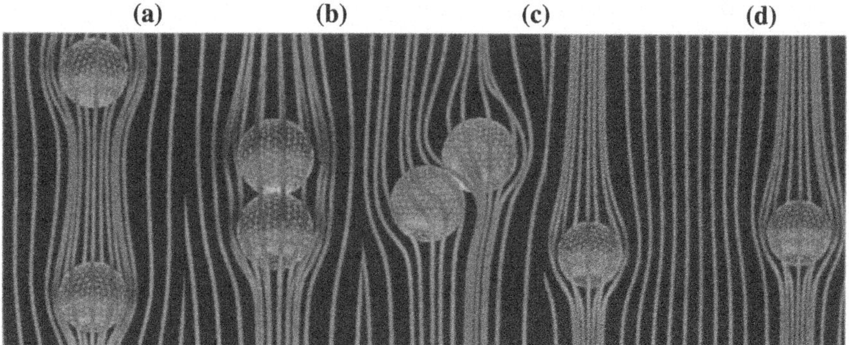

Fig. 14 Interaction of a sedimenting pair of particles - drafting (**a**), kissing (**b**), tumbling (**c**–**d**) (Prosperetti and Tryggvason 2007)

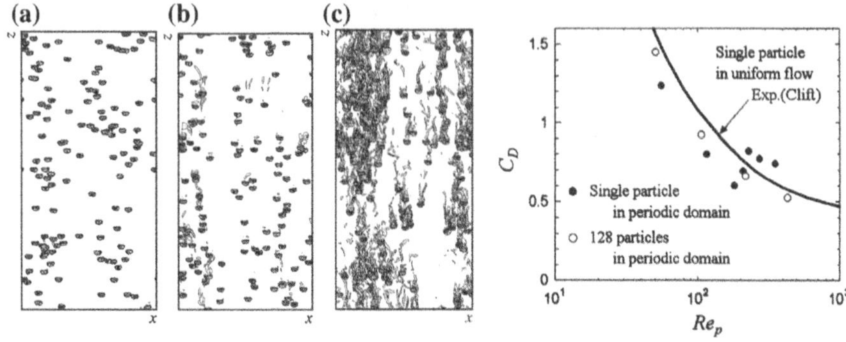

Fig. 15 *Left* Instantaneous flow field including 128 particles for particle Reynolds numbers 100, 200, and 400 (from *left* to *right*). *Right* Drag coefficient of an individual particle and a sedimenting cloud of 128 particles (Kajishima and Takiguchi 2002)

5.2 Sedimentation

The last paragraph introduced the important effect of wake attraction. We can expect that it will also play an important role when many particles sediment under gravity. This situation has been studied by Kajishima and Takiguchi (2002) who performed DNS of 128 identical particles (same size and density) sedimenting under the action of gravity. They found two remarkable finite-size effects: First, wake attraction leads to the formation of clusters (see Fig. 15 (left)). Trailing particles are slip-streaming and thus approaching the leading particles. In principle, this works for many particles as for two particles. In consequence particles agglomerate in vertically elongated clouds. The stability and evolution of such clusters with respect to the intrinsically generated velocity fluctuations is a more complicated and still an open question. The second finite-size effect is that, the average drag force experienced by sedimenting particles is reduced if many particles are present (see Fig. 15 (right)). The wake attraction effect gives thus rise to a collective effect in reducing the force on the individual particles.

This drag reduction is also reflected in the sedimentation speed. Uhlmann (2005) considered the evolution of the average vertical velocity of particles initially at rest. He performed several DNS and varied the number of particles. A single particle accelerates monotonously and reaches asymptotically its terminal velocity. Uhlmann (2005) showed that a set of particles with initial positions on a regular grid passes through different stages. Once released they accelerate as expected (see Fig. 16 (right)). Soon they reach (on average) a higher velocity than a single particle. During this phase, the wake attraction effect reduces the drag and allows for higher vertical velocities. However, once wake induced fluctuations make the particle deviate from their initially regular spatial distribution their settling velocity reduces. It reduces to values smaller than that of a single particle. The important point here is that the particle Reynolds number ($Re_p \approx 400$) is sufficiently high that the particle wakes

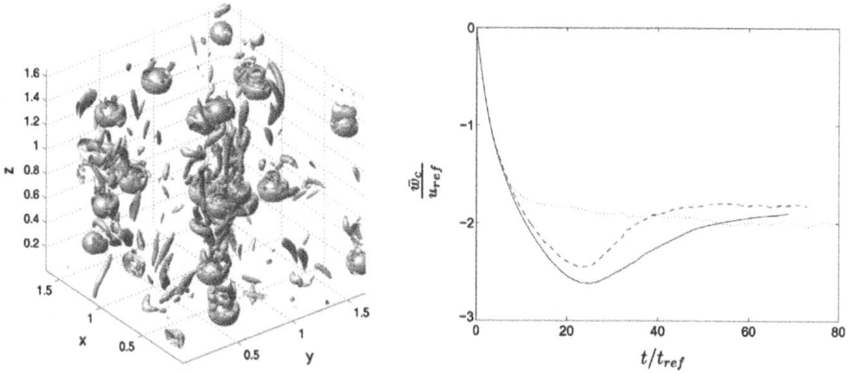

Fig. 16 *Left* Instantaneous flow field including $Re_p \approx 400$ particles. *Right* Average settling speed as a function of time (Uhlmann 2005)

are turbulent (see Fig. 16 (left)). The particle-wake interactions lead to an increased drag and decreased settling velocity. Uhlmann (2005) shows that a higher particle number leads to a lower average settling speed. This is in agreement with the observation (from Sect. 4.3) that the drag force increases with increasing turbulent intensity because a higher number of particles leads to more fluctuations (see Fig. 16 (left)) and in turn to a higher turbulent intensity.

In a later numerical work Uhlmann and Doychev (2014) focused on possible clusters of many sedimenting particles. They considered two different systems having the same solid to fluid density ration of 1.5 and the same solid volume fraction of

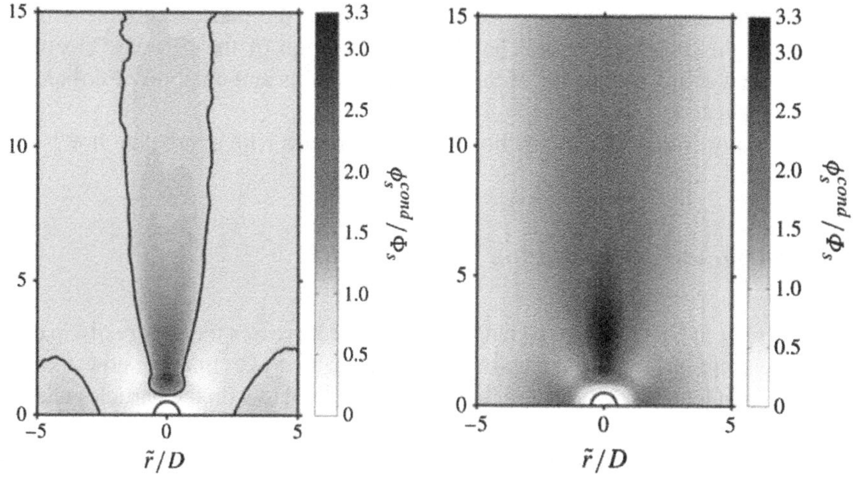

Fig. 17 Averaged solid volume fraction conditioned on particle positions, normalized by the global solid volume fraction. *Left* $Ga = 121$. *Right* $Ga = 178$ (Uhlmann and Doychev 2014)

0.005 but different Galileo numbers $Ga = u_g D/\nu, u_g = \sqrt{|\rho_p/\rho_f - 1|Dg}$. Starting from initially random particle positions, they found that the particles do not always agglomerate. Only the system with the higher Galileo number ($Ga = 178$) showed significant particle concentrations while the simulation with $G = 121$ did not. This is visualized in Fig. 17 where the normalized average solid volume fraction conditioned on the particle position is shown. This quantity is the lateral-angle averaged volume fraction of particles surrounding a given particle in the cylindrical coordinate system $((\tilde{r}, \tilde{z})$ = (lateral,vertical)). It is normalized to the mean solid volume fraction. This representation reveals important features about the average particle distribution seen by individual particles. The leading particle often pulls a trailing particle very close to it manifested by the dark disc at a distance of approximately one diameter. In the lower Ga case (Fig. 17, left) this disc has a short shadow indicating the presence of second trailing particle. In the higher Galileo number case (Fig. 17, right) this shadow is not only much longer but also broader showing the coordinated movement of many particles and thus a cluster. The existence of the latter can also be concluded from the lateral density profile. For $Ga = 121$ one does not observe an increased particle density at the sides of the reference particle while this is the case for $Ga = 178$.

Their explication of these Galileo number based differences is based on the single particle dynamics: Above a critical value of $Ga = 155$ the trajectory of a sedimenting particle undergoes an instability and becomes oblique (see Sect. 4.4). They claim that this enhanced span-wise mobility of the particles compared to that of vertically sedimenting particles facilitates the occurrence of clusters. The idea is thus that many particles moving on random oriented oblique paths have better chances the encounter the wakes of other particles than particles falling simply vertically. The end of the story is here again the wake attraction effect which make the particles catch up each other.

The existence of these clusters has important effects on the dynamics of the particles and also on that of the fluid. For instance, the particle motion is more agitated in the clustering case. The standard deviation of the particle velocity is higher which is a consequence of the wake interactions and creation of turbulence within the clusters.

We will now turn to the interaction of many particles with a turbulent flow.

5.3 Turbulence Modulation

We have seen in Sect. 4.2 that an individual particle increases the energy dissipation rate close to its surface while it calms turbulence in its wake. Imagine now, that we seed our turbulent flow with many finite-size particles. How do the particles change the turbulent flow? Do they increase or decrease the kinetic energy dissipation rate?

Cate et al. (2004) studied these questions by seeding a forced turbulent flow with a few thousand finite-size particles that were slightly heavier than the fluid. The volume fraction (the volume occupied by the particles compared to the total volume of the fluid) varied from 2 to 10%. They found that the spatial distribution of energy

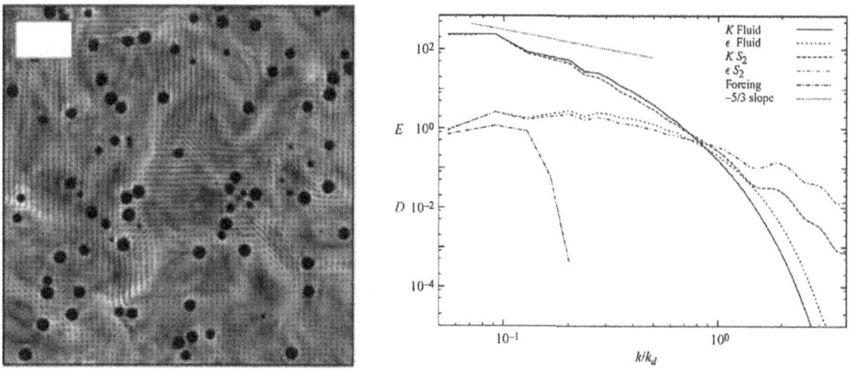

Fig. 18 *Left* Instantaneous local energy dissipation. *Right* Spectrum of energy dissipation (Cate et al. 2004)

dissipation is significantly influenced by the particles (see Fig. 18 (left)). Beside their boundary layer they also create inter-particle structures of increased dissipation. This could be a real collective effect in the sense that is it might be different from a simple superposition of the imprints of individual particles.

Changes in the energy dissipation are also clearly reflected in the global dissipation spectrum (see Fig. 18 (right)). Compared to a single phase flow (without particles) the dissipation rate is reduced at scales larger than the particle diameter while it is increased a smaller scales. Particles thus reduce shear at large scales while they increase it at small scales. The authors find the analog for the scale dependent kinetic energy. Particles transfer energy from large scales to scales of the order of the particle diameter or smaller.

There is another interesting result in the work of Cate et al. (2004). They computed the characteristic collision time scale of binary collisions and compared it to a prediction for small particles in a turbulent flow. Their data agrees with the prediction for small volume fraction of 2% and start to deviate at 5%. In fact, their collision times are longer than what one expects which in turn means that collective effects might reduce the probability of collisions.

Lucci et al. (2010) studied a case similar to the one discussed before. The main difference is that their flow is not forced so that they investigate the influence of many finite-size particles on decaying turbulence. They also find that the particles increase the energy dissipation rate and that they transfer energy from large to smaller scales. Furthermore, they find that the two-way coupling rate $\Psi_p(t) = \langle u_i f_i \rangle$ where f is the force exerted by the particle on the fluid is always positive. In decaying turbulence finite-size particle act as a source of kinetic energy while $\Psi_p(t)$ can be positive or negative for small particles.

Yeo et al. (2010) compared the dynamics of many finite-size particles in forced turbulence for particles that were heavier or lighter than the fluid. They found that heavy particles are expelled from vortex filaments while light particles are entrapped in these structures, an observation also made for point-particles. Concerning the

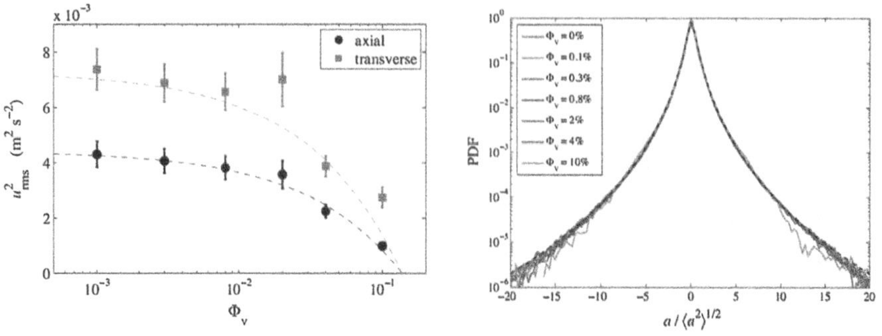

Fig. 19 *Left* Variance of the turbulent velocity fluctuations u^2_{rms} as a function of the particle volume fraction; The *dashed lines* are $u^2_{rms}(\Phi_v) = u^2_{rms}(0)(13.75\,\Phi_V)$. *Right* Normalized probability density functions of acceleration components for various values of the volume fraction Φ_V as labeled (Cisse et al. 2015)

energy dissipation spectrum they confirm the previously discussed results. Interestingly, they found that the modulation of turbulence only weakly depends on the type of particles (light or heavy) but is dominated by the finite-size contributions to the flow disturbances.

So far, there are very few experimental investigations on turbulence modulation by finite-size particles. The reason are mostly technical difficulties. In a water experiment Bellani et al. (2012) studied a turbulent suspension of hydro-gel particles by means of a particle-image velocimetry technique. Hydro-gel is nearly neutrally buoyant and has a refraction index close to that of water. They found that for a volume fraction of $1.3 \cdot 10^{-3}$ the turbulent kinetic energy is reduced by 14% compared to an unladen flow.

Recently, Cisse et al. (2015) optically tracked the movement of tracer-like particles transported by a von Kármán flow. For their experiments they seeded water with a varying number of finite-size particles. For the latter they choose super-absorbent polymer spheres whose optical index and mass density match those of water. From the tracer dynamics they deduced that the big particles globally attenuate the fluid turbulence. When increasing the particle occupied volume Φ_V the turbulent velocity fluctuations reduces as $\Phi_V^{2/3}$ (see Fig. 19 (left)). It is not surprising that the level of turbulence decreases because this has already been observed for an individual particle (see Sect. 4.2). What is surprising is the exponent 2/3. The authors interpret the latter by the possibility that not all added particles (their total volume) contribute to the turbulence reduction but only a fraction as if particles agglomerate on a surface rather than distribute in the full volume of the flow. Unfortunately, the authors did not track the large particles to conclude on this hypothesis. The evoked clustering is a possible collective effect and needs further investigation especially because Fiabane et al. (2012) have found that neutrally buoyant finite-size particles do not cluster in homogeneous isotropic flows even if their Stokes number is large. But neutrally buoyant particle clustering can occur in non-homogeneous flows: Machicoane et al.

(2014) found clear evidence of preferential concentrations in a turbulent von Kármán flow. Particles tend to stay in the vicinity of the rotating impellers and avoid the central region of the device.

Not all statistics of the flow are affected by the presence of the finite-size particles. Following Cisse et al. (2015), it seems that higher-order small-scale quantities are insensitive to the addition of big particles. The normalized acceleration probability density functions fall on top of each other irrespective of the volume fraction Φ_V (see Fig. 19 (right)).

6 Concluding Remarks

Let us summarize the most important findings that distinguish finite-size from point particles.

We first begin with results for individual particles:

- *Point particles* have (by definition) a negligible Reynolds number and are much smaller diameter than any characteristic scale of the flow. Their surrounding flow profile (Stokes flow) and drag force (Stokes drag) can be computed analytically.
- *Finite-size particles* are surrounded by a more complicated boundary layer and wake structure. In general, no analytic expression is known for the force on such a particle.
  - Their numerical treatment is demanding and various different strategies are proposed to integrate the Navier–Stokes equations together with the no-slip boundary condition at the particles surface.
  - The very definition of a slip velocity with respect to the fluid and consequential particle Reynolds number is difficult.
  - The energy dissipation rate is increased very close the particle while it can be reduced in the wake.
  - Large and small scale turbulent fluctuations lead to an increase of the drag force on the particle.
  - Particles with even steady wakes can sediment under gravity on non-vertical paths.

Let us now list collective effects arising from the mentioned finite-size effects:

- A particle entrains fluid in its wake that can reduce the drag force on a trailing particle. This is called the wake attraction effect.
- Particles separated in cross-stream direction are repulsed from each other at very small distances and are attracted at larger distances.
- The wake attraction effect leads to the formation of particle clusters.
- Clusters sediment on average at higher speeds than individual particles.
- Finite-size particles transfer turbulent kinetic energy from scales of the order of the particle diameter to smaller scales.
- Collective hydrodynamic interactions might reduce the collision rate.

In summary one can conclude that a better understanding of finite-size particle suspension will need much more analytic, numerical and experimental efforts.

Finally, we should mention that the Reynolds number of the presented works are quite limited and will not allow for a clear extension of the inertial range of scales. The results will probably include finite-Reynolds number effects and might thus change for higher Reynolds numbers. This kind of question has also to be answered in the future.

References

P. Bagchi and S. Balachandar. Effect of turbulence on the drag and lift of a particle. *Phys. Fluids*, 15(11):3496, 2003. doi:10.1063/1.1616031.

G. Bellani and E.A. Variano. Slip velocity of large neutrally buoyant particles in turbulent flows. *New J. Phys.*, 14:125009, 2012.

G. Bellani, M.L. Byron, A.G. Collignon, C.R. Meyer, and E.A. Variano. Shape effects on turbulent modulation by large nearly neutrally buoyant particles. *J. Fluid Mech.*, 712:41–60, 2012.

R. Bitane, H. Homann, and J. Bec. Geometry and violent events in turbulent pair dispersion. *Journal of Turbulence*, 14:23–45, 2013.

T. Burton and J. Eaton. Fully resolved simulations of particle-turbulence interaction. *J. Fluid Mech.*, 545:67–111, 2005a.

T. Burton and J. K. Eaton. Fully resolved simulations of particle-turbulence interaction. *J. Fluid Mech.*, 545:67, 2005b.

M. Cisse, H. Homann, and J. Bec. Slipping motion of large neutrally buoyant particles in turbulence. *Journal of Fluid Mechanics*, 735:R1, 2013.

M. Cisse, E.-W. Saw, M. Gibert, E. Bodenschatz, and J. Bec. Turbulence attenuation by large neutrally buoyant particles. *Physics of Fluids*, 27:061702, 2015.

A. Daitche. On the role of the history force for inertial particles in turbulence. arXiv:1501.04770, 2015.

E A Fadlun, R Verzicco, P Orlandi, and J Mohd-Yusof. Combined Immersed-Boundary Finite-Difference Methods for Three-Dimensional Complex Flow Simulations. *J. Comp. Phys.*, 60:35–60, 2000.

L. Fiabane, R. Zimmermann, R. Volk, J.-F. Pinton, and M. Bourgoin. Clustering of finite-size particles in turbulence. *Phys. Rev. E*, 86:035301, 2012.

R. Gatignol. The Faxén formulae for a rigid sphere in an unsteady non-uniform Stokes flow. *J. Méc. Théor. Appl.*, 1:143–160, 1983.

H. Homann, J. Bec, and R. Grauer. Effect of turbulent fluctuations on the drag and lift forces on a towed sphere and its boundary layer. *J. Fluid Mech.*, 721:155–179, 2013.

T. Kajishima and S. Takiguchi. Interaction between particle clusters and particle-induced turbulence. *International Journal of Heat and Fluid Flow*, 23:639–646, 2002.

A. Kidanemariam, C. Chan-Braun, T. Doychev, and M. Uhlmann. Direct numerical simulation of horizontal open channel flow with finite-size, heavy particles at low solid volume fraction. *New J. Phys.*, 15:025031, 2013.

J. Kim and S. Balachandar. Mean and fluctuating components of drag and lift forces on an isolated finite-sized particle in turbulence. *Theoretical and Computational Fluid Dynamics*, 26:185–204, 2012.

I. Kim, S. Elghobashi, and W. A. Sirignano. Three-dimensional flow over two spheres placed side by side. *Journal of Fluid Mechanics*, 246:465, 2006.

F. Lucci, A. Ferrante, and S. Elghobashi. Modulation of isotropic turbulence by particles of taylor length-scale size. *J. Fluid Mech.*, 650:55, 2010.

N. Machicoane, R. Zimmermann, L. Fiabane, M. Bourgoin, J.-F. Pinton, and R. Volk. Large sphere motion in a nonhomogeneous turbulent flow. *New Journ. Phys.*, 16:013053, 2014.

M. R. Maxey and B. K. Patel. Localized force representations for particles sedimenting in stokes flow. *International Journal of Multiphase Flow*, 27:1603–1626, 2001.

M. R. Maxey and J. Riley. Equation of motion for a small rigid sphere in a nonuniform flow. *Phys. Fluids*, 26(4):883, 1983.

A. Naso and A. Prosperetti. The interaction between a solid particle and a turbulent flow. *New Journ. Phys.*, 12(3):033040, 2010. ISSN 1367-2630.

A. Prosperetti and H.N. Oguz. Physalis: A New o(N) Method for the Numerical Simulation of Disperse Systems: Potential Flow of Spheres. *J. Comp. Phys.*, 167:196–216, 2001.

A. Prosperetti and G. Tryggvason. *Computational methods for multiphase flow*. Cambridge University Press, 2007.

L. Schiller and A. Naumann. über die grundlegenden berechnungen bei der schwerkraftaufbereitung. *Verein Deutscher Ingenieure*, 77:318, 1933.

A. Ten Cate, J. Derksen, L. M.Portela, and H. E. Van Den Akker. Fully resolved simulations of colliding monodisperse spheres in forced isotropic turbulence. *Journal of Fluid Mechanics*, 519:233–271, 2004.

M Uhlmann. An immersed boundary method with direct forcing for the simulation of particulate flows. *J. Comp. Phys.*, 209(2):448–476, 2005.

M. Uhlmann and T. Doychev. Sedimentation of a dilute suspension of rigid spheres at intermediate galileo numbers: the effect of clustering upon the particle motion. *Journal of Fluid Mechanics*, 752:310–348, 2014.

M. Uhlmann and J. Dušek. The motion of a single heavy sphere in ambient fluid: A benchmark for interface-resolved particulate flow simulations with significant relative velocities. *International Journal of Multiphase Flow*, 59:221–243, 2014.

K. Yeo, S. Dong, E. Climent, and M.R. Maxey. Modulation of homogeneous turbulence seeded with finite size bubbles or particles. *International Journal of Multiphase Flow*, 36:221–233, 2010.

C. Zhu, S. C. Liang, and L. S. Fan. Particle wake effects on the drag force of an interactive particle. *International journal of multiphase flow*, 20(1), 1994.

Some Aspects of the Collective Dynamics of Particles in Turbulent Flows

Mickaël Bourgoin

Abstract This chapter presents some important features of collective dynamics of particles transported by a turbulent flow. The focus is on two main processes. The first concerns the dispersion of fluid tracers (i.e. particles which follow exactly the flow) and revisits the classical problem of turbulent pair dispersion. It is shown, that a simple scale-dependent iterative ballistic process accurately accounts for most turbulent super-diffusive properties, which are explicitly related to the multi-scale nature of turbulence via the turbulent energy spectrum. The second considers the case of inertial particles (which are not tracers of the carrier flow) and the phenomenon of preferential concentration, responsible for the formation of persistent particle clusters. Tools to diagnose the presence of clustering are reviewed, together with the main properties of clusters in turbulence and of the underlying particles/turbulence interaction mechanisms.

1 Introduction

Particle laden turbulent flows are ubiquitous, both in natural systems (sediments, plankton, aerosols, cloud droplets and rain, etc.) and human activities (sprays, powders, combustion, etc.). Predicting the dynamics of an ensemble of material particles dispersed and transported in a turbulent environment remains a major problem which has motivated countless studies over more than a century. Among the many situations where turbulent transport of particles is important, environmental issues are worth being emphasized. The low atmosphere carries in average more than a thousand particles per cubic centimeter (so called aerosols) with diameter ranging from several nanometers to hundreds of micrometers. Larger (millimetric) particles, such as rain droplets, solid debris, etc., can also be present. The global cycle of these particles is very complex. Their origin is both natural (sand, ashes, marin salt, iberolites, pollens, condensation of drops and droplets, etc.) and anthropogenic

M. Bourgoin (✉)
Physics Laboratory, Ecole Normale Superieure of Lyon, Lyon, France
e-mail: mickael.bourgoin@ens-lyon.fr

C. Marchioli (ed.), *Collective Dynamics of Particles*,
CISM International Centre for Mechanical Sciences 576,
DOI 10.1007/978-3-319-51226-6_3

(industrial pollutants, artificial radionuclides, etc.). Larger particles sediment locally. Smaller particles are transported accross the atmosphere (where they can undergo several physical and chemical transformations) at planetary scale until they are eventually deposited. While in suspension, they play a crucial role on many atmospheric phenomena (nucleation sites for cloud droplets, rain, scattering and absorption of solar and terrestrial radiation, oxidant capacity of the atmosphere, etc.). When deposited, their impact is also important with possible positive effects (as pollination) but also eventual harmful consequences (change of albedo of large glaciers, water contamination, deposition of adsorbed pollutants, respiratory health hazard, etc.). In a global planetary balance, the atmospheric transport and deposition of particles therefore affect (directly and indirectly) Earth climate and human health among other important contemporary challenges. Understanding the turbulent transport of particles is therefore a topic of primary importance.

One of the main difficulties lies on the intrinsically random and multi-scale (in space and time) nature of turbulence with which transported particles interact. Turbulence is indeed characterized by a hierarchy of random motion with a wide range of dynamical scales (called *inertial range of turbulence*) from the largest (and slowest) eddies where energy is injected down to the smallest (and fastest) at which it is dissipated by molecular viscosity. The extent of the inertial range is directly related to the Reynolds number Re of the carrier turbulence ($Re = u_{rms}L/\nu \propto (L/\eta)^{4/3} \propto (T_L/\tau_\eta)^2$, where u_{rms} is the turbulent fluctuating velocity, L and T_L and the energy injection length and time scale and η and τ_η the dissipative length and time scale). Despite its ubiquitousness, turbulence remains one of the deepest unsolved mysteries of classical physics. Even if an exact theory for turbulence does a priori exist, the strong non-linearities of Navies-Stokes equations have vanished so far all attempts to find an analytical solution of the problem. We are therefore committed to find the best possible description of the phenomenon in terms of an enumeration (which ought to be as complete as possible) of the mathematical and physical properties mainly in a statistical sense of these unknown solutions, seeking in particular for universal behaviors. The first stone in building such a statistical description in modern history of turbulence starts with Richardson, who proposed in the 1920s a multi-scale description of turbulence in terms of an energy cascade, where turbulence appears as a hierarchy of random eddies with sizes ranging from the scale L where energy is injected down to the scale η where it is dissipated by viscosity. This range of scales defines the *inertial range* of turbulence. Though we know today that this cascade results from the non-linear interaction of Fourier modes of the velocity field, we are still unable to quantitatively model and predict turbulence statistics over the entire range of inertial scales. Since 1941, Kolmogorovs ideas (refined in 1962 by Obukhov) dominate the common way to describe turbulence. Taken together, Kolmogorov's hypotheses state that in intense turbulence and well away from any boundaries or singularities the statistics of turbulent flow should be universal at length and timescales that are small compared with the injection of energy into the flow. In Kolmogorov's approach the mean energy dissipation rate ϵ then becomes the only relevant parameter governing the dynamics of structures in the inertial range. Since then, statistical turbulence modeling has been dominated by Kolmogorov's ideas, whose 1941 hypotheses have so influenced the field that they are simply known as the K41 phenomenology. The great utility of the K41 model lies in its prediction of universal scaling laws (most of which simply reflect dimensional constraints within K41 hypothesis) for velocity increments statistics (so called *structure funcitons*). Kolmogorov's phenomenology remains the dominant angle of attack of turbulence research. In spite of some successes (as the 5/3 turbulent spectrum, the 4/5 law or recent progresses in multi-fractal description) it is however a constative fact that our toolbox of concepts for turbulence investigation remains incomplete and not suited yet to produce a full understanding of the phenomenon (finite Reynolds number effects, anisotropy effects, intermittency phenomenon, turbulent transport, etc. are, for instance, just a few of the remaining mysteries). Numerical simulations and experimental observations have nevertheless reached today a high level of accuracy regarding the characterization of random multi-scale properties of turbulence.

1.1 Particles in Turbulence

The situation becomes even more complex when particles are added to turbulent flows. Depending on their size and density relative to the fluid (eventually responsible for a finite response time of the particle due to its inertia), particles will interact with structures of the carrier flow at different time and spatial scales.

Tracer particles. Neutrally buoyant particles much smaller than the dissipation scale η are expected to behave as actual tracers of fluid particles, as they are capable to respond to the smallest and fastest structures of the carrier flow (such particles are comonly used in experiments to characterize the flow itself) and their motion is ruled by Navier-Stokes equations. The investigation of the turbulent dynamics and trajectories of such particles is known as the Lagrangian description of turbulence. It has become in the very last decade one of the most accurate methods to characterize experimentally the multi-scale statistics of turbulence. This is to be related to the explosive and constant improvements of high resolution and high-speed digital cameras (as well as other alternative techniques, including acoustical tracking, extended laser doppler velocimetry and instrumented particles), which has opened a new era in fluid mechanics metrology, a situation which we could call the *Lagrangian Revolution* (Bourgoin et al. 2014), emphasizing at the same time the renewed interest for the Lagrangian paradigm (Toschi and Bodenschatz 2009) (with respect to the usual Eulerian approach, where the flow is described in terms of fields rather than trajectories) and the fact that Lagrangian experimental methods are still at their infancy, but ready to become mature. High-resolution (in space and time) 3D Lagrangian particle tracking systems are now capable to track thousands of particles simultaneously in highly turbulent flows (Bourgoin et al. 2006).

Inertial particles. On the contrary particles with density mismatch and/or with size comparable to inertial turbulent eddies, do not follow the flow exactly. Such particles (large and/or density mismatched) are generically referred to as *inertial particles*. The exact physical mechanisms coupling the dynamics of such particles and turbulence remains an open question and a very active field of research. Several inertial effects have been identified for long, and are commonly interpreted in qualitative terms of particles interacting with turbulent eddies. For instance large particles are expected not to respond to turbulent eddies much smaller than their own size (what results in a spatial filtering of turbulent fluctuations from the carrier flow); similarly, density mismatch induces a finite response time of the particle (due to inertia), which inhibits the impact of the fastest fluctuations of the carrier flow (what results in a temporal filtering of turbulent fluctuations). However, an accurate quantitative description of such effects is still lacking. One of the reasons is our inability to write a generic and proper equation of motion for inertial particles in a turbulent environment. We are therefore in the situation where turbulence by itself is a difficult problem, but for which we do have a master equation, the Navier-Stokes equations (although we do not have a solution), while for particles we do not even know how to write an appropriate equation of motion accounting for the unsteady forces the flow exerces on the particles. The formulation of the problem is however straightforward: considering a rigid dense sphere in a viscous newtonian flow verifying Navier-Stokes equations, with no-slip boundary conditions on the sphere surface, can we formalise the instantaneous action of the flow on the sphere? The best answer we still have to this longstanding question was formulated successively by Basset (1888), and later refined by Boussinesq (1903) and Oseen (1911), who examined the motion of a sphere settling under gravity in a fluid at rest. Their analysis started from the simplest unsteadiness situation of a sphere settling from rest in a quiescent viscous fluid, which concerns the transient acceleration until it reaches its terminal settling regime. The description of this simple phenomena already includes important generic ingredients and complexities of the phenomenology of particle-flow interactions in all classes of unsteady flows (such as turbulence). The disturbance flow produced by the motion of the sphere was assumed to have sufficiently low Reynolds number so that the fluid force on the sphere could be calculated from Stokes flow. Tchen extended this work to a sphere settling under gravity in an unsteady and nonuniform flow, with a view already to turbulent flows. The resulting model (known as BBOT, Basset-Boussinesq-Oseen-Tchen) has been revisited in 1983, simultaneously by Maxey and Riley (1983) and Gatignol (1983), leading to the following formulation for the equation of motion for a particle in a flow:

$$m_p \frac{d\vec{v}}{dt} = 3\pi\mu_f d_p(\vec{u} - \vec{v}) + \frac{1}{2} m_f \frac{d(\vec{u} - \vec{v})}{dt} + m_f \frac{D\vec{u}}{Dt} + \frac{3}{2} d_p^2 \sqrt{\pi \rho_f \mu_f} \int_{-\infty}^{t} \frac{d(\vec{u} - \vec{v})}{dt} \frac{d\tau}{\sqrt{t - \tau}} + (m_p - m_f)\vec{g}, \quad (1)$$

where $\vec{v}$ is the particle velocity, $\vec{u}$ is the unperturbed carrier flow velocity field, d_p is the particle diameter, ρ_p is the particle density, ρ_f is the carrier fluid density, $m_f = \rho_f \pi \frac{d_p^3}{6}$ is the mass of the fluid displaced by the particle and μ_f is the carrier fluid dynamic viscosity. The terms on the right hand side are, in order of appearance: (i) the Stokes drag force (due to the relative velocities of particle and fluid), (ii) the added mass force, which is purely inertial, and corresponds to the force exerted by the displaced fluid, (iii) the pressure gradient term, which is equivalent to the fluid particle acceleration at the position of the particle center, (iv) the history term, which takes into account the entire history of the particle motion in the carrier fluid up to an instant t and mainly takes into account the interaction of the particle with its own wake and (v) the Archimedes force. The actual range of validity of this equation is however limited as it is only valid for vanishing particulate Reynolds number $Re_p = d_p|\vec{v} - \vec{u}|/\nu \ll 1$ (as it assumes a Stokes flow around the particle due to the relative velocity $\vec{v} - \vec{u}$ between the particle and the fluid) and for particles much smaller than the local non-uniformities of the flow (as it supposes that the unperturbed flow velocity $\vec{u}$ can be uniquely defined at the particle position). It is therefore generally referred to as a *point particle model*. In spite of this strong limitation, apart from some first order corrections detailed below to account for finite Re_p effects and local non-uniformities of the flow around the particle (so called Faxen corrections), the point particle model still remains the most achieved analytical expression we have at hands to describe particles-fluid interaction. In its simplest implementation, the equation of motion (1) is primarily dominated by the drag force induced by the relative motion between the particle and the carrier flow. This simplification offers a tractable theoretical framework with an abundant dedicated litterature (Toschi and Bodenschatz 2009) (although with an unclear range of validity), where particles velocity is simply related to the fluid's by the relation

$$\frac{d\vec{v}}{dt} = \tau_p^{-1}(\vec{u} - \vec{v}). \tag{2}$$

Particle inertia is then quantified by a single dimensionless number, namely the Stokes number St, which compares the viscous relaxation time of the particle $\tau_p = d_p^2(1 + 2\Gamma)/36\nu$ (assuming a Stokes flow around the particle) to a characteristic time scale of the carrier flow τ_f (generally taken as τ_η, the dissipative time-scale of turbulence): $St = \tau_p/\tau_\eta = (d_p/\eta)^2(1 + 2\Gamma)/36$ (recalling that temporal and spatial dissipative scales are related to the kinematic viscosity by $\nu = \eta^2\tau_p^{-1}$). We shall refer to this simplification as the *Stokesian model*. Note that this model assumes both $d_p \ll \eta$ and $Re_p = d_p|\vec{v} - \vec{u}|/\nu \ll 1$. The first ensures that the velocity field is smooth and well-defined at the particle scale, hence allowing for instance to define *the velocity of the fluid at the particle position*, whereas the second approximation ensures that the relative velocity between the particle and the fluid remains small enough to assume a linear drag.

1.2 Some Important Aspects of the Collective Dynamics of Particles in Turbulence

When many particles are transported simultaneously by the flow, they collective dynamics is intimately related to the way they interact with the carrier turbulence. The problem is intrinsically complex, as the random and multi-scale nature of turbulence induces a non-trivial collective motion of the particles, even for the case of passive particles in *one-way coupling* configuration, where the volume fraction is low enough so that the presence of the particles do not modify the flow and particles do not interact with each other. The situation is naturally even more intricate as the volume fraction increases and *two-way coupling* effects (back reaction of the particles on the carrier flow) and *four-way coupling* effects (collisions between particles) emerge. We will focus here only on the *one-way* coupling situation, where particles collective dynamics is entirely controlled by the sole action of the turbulence on the particles. Note that, although this is by essence a Lagrangian problem, where the turbulent dynamics of the transported inclusions is naturally addressed in terms of particles trajectories, it is also intimately related to mixing properties of turbulence, which is generally addressed as a Eulerian problem, in terms for instance of the concentration field of a dispersed scalar. In the course of its long history, the study of turbulent flows always benefited from comings and goings between fields and particles, that is between an Eulerian description and a Lagrangian approach. Pioneered in the 1920s by the G.I. Taylor (1922) and L.F. Richardson (1926), the Lagrangian statistical formalism (where the flow is described in terms of the motion of fluid particles along their trajectories) suffered from a lack of reliable and precise experimental measurements. It is only during the last decades that new highly sophisticated laboratory experiments (Ott and Mann 2000; Voth et al. 2002; LaPorta et al. 2001; Bourgoin et al. 2006, 2014) (allowing to track particles in highly turbulent flows with sufficient spatial and temporal resolution) and high-resolution direct numerical simulations (Yeung and Pope 1989; Biferale et al. 2005) have attacked seriously the problem of Lagrangian turbulence, and more widely, the problem of the transport of impurities (not restricted to infinitesimal material fluid elements) by a turbulent flow (the reader can refer to the following reviews (Toschi and Bodenschatz 2009; Bourgoin et al. 2014) for more details on recent experimental, numerical and theoretical advances regarding the Lagrangian approach of turbulent transport). This revival has led to new great challenges regarding the collective motion of particles transported in a turbulent flow, that are nowadays on the ground. The present chapter focuses on two main aspects of the turbulent transport of particles: (i) the turbulent dispersion of tracer particles and (ii) the turbulent dispersion of inertial particles. In both situations the interaction of the particles with the multi-scale properties of the carrier turbulence can result in strong heterogeneities of the concentration field of particles. From a practical point of view, this phenomenon can play a crucial role in many situations. Particle laden flows are indeed ubiquitous, both in natural and anthropogenic flows, an emblematic example being the atmospheric transport of particles and pollutants.

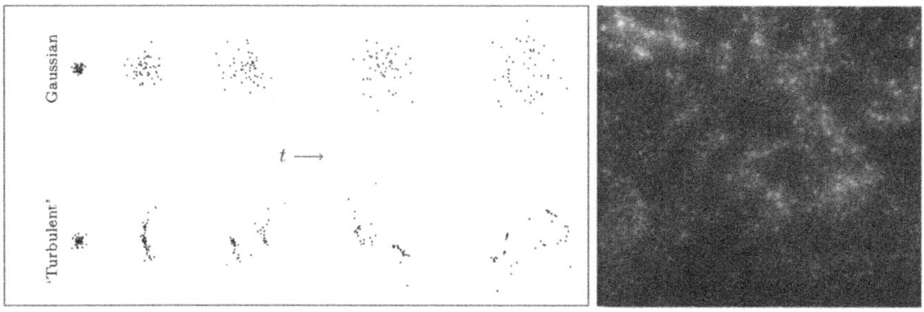

Fig. 1 **a** Sketch of the dispersion of two puffs with different relative dispersion statistics (image taken from Mann et al. 1999) (*top* brownian dispersion, *bottom* superdiffusive dispersion). **b** Evidence of clustering of water droplets in a homogeneous isotropic turbulent flow of air (image taken from Wood et al. 2005)

The variance of fluctuations of the scalar concentration of a substance dispersed from a source point can for instance be related to the relative dispersion of pairs of fluid tracers (NELKIN and KERR 1981; Sawford 2001). While a classical brownian process will lead to a smooth gaussian diffusion, turbulence is known to behave superdiffusively. To illustrate this, Fig. 1a shows schematically the dispersion of a cloud of particles, assuming in one case a classical brownian dispersion and in the other case a superdiffusive dispersion. Superdiffusion produces highly intermittent particles distributions (which in a Eulerian approach would lead to an intermittent concentration of a dispersed scalar field) with highly concentrated regions surrounded by depleted zones. Though this superdiffusivity is known since the pioneering work by Richardson in (1926), despite more than 80 years of scientific inquiry into relative dispersion (Huber et al. 2001; Sawford 2001; Richardson 1926; Batchelor 1952; Ott and Mann 2000; Biferale et al. 2005; Bourgoin et al. 2006), no clear experimental verification of the theoretical predictions has emerged yet and it is only recently that numerical simulations have converged toward a relatively well characterized description of the process. Section 2 of this chapter proposes a survey on the relative dispersion problem.

The situation is further complexified when the dispersed particles are not a fluid tracer, but material particles, such as dust, droplets, aerosols, etc. As they disperse, such small impurities (whose dynamics is not reducible to that of passive tracers) can develop stronger concentration events where fractal clustering emerges even in incompressible flow and even if a homogeneous random initial distribution of particles is prescribed (Fig. 1b). This peculiar behavior, known as the *preferential concentration* phenomenon, is directly related to the dissipative nature of the dynamics of inertial particles (primarily related to the drag induced by the particle-fluid relative velocity, as emphasized in Stokesian models such as Eq. 2) and to the action of the multi-scale turbulent eddies on the particles. A simple intuition of this phenomenon can be understood for instance in terms of centrifugal expulsion of denser particles out of the core of turbulent eddies. Heavy particles are therefore expected to preferentially cluster outside turbulent eddies. On the contrary particles lighter than the carrier fluid will tend to preferentially the core of turbulent eddies. At the experimental level such inhomogeneities have been known for a long time (see Toschi and Bodenschatz 2009 for a review) and used for flow visualization (e.g. exploiting bubble clustering inside vortex filaments). Section 3 of this chapter will be devoted to the preferential concentration phenomenon.

2 Turbulent Dispersion of Tracer Particles

2.1 The Turbulent Pair Dispersion Problem

Molecules in a quiescent fluid spread due to molecular diffusion. If we consider a small spherical patch of tagged molecules, this results in an isotropic and homogeneous growth of the patch. At a microscopic level this expansion is due to random uncorrelated collisions induced by the thermal agitation of the molecules. At a macroscopic level this mechanism results in a Fickian diffusion process where the local concentration C of tagged molecules diffuses according to the simple equation $\partial C/\partial t = K\Delta C$, where K is the molecular diffusivity, with units $[\mathrm{m}^2 \cdot \mathrm{s}^{-1}]$. In elementary kinetic gas theory, the connection between microscopic and macroscopic descriptions is for instance given by the relation $K \propto l v_T$ (with l a characteristic correlation length of particles trajectories, typically given by the mean free path and v_T the thermal agitation velocity of the molecules). A fundamental property of such a Fickian process concerns the linear growth with time t of the mean square separation $< \vec{D}^2 > \propto Kt$ between any two molecules in the patch, what is generally referred to as *normal diffusion*. Normal molecular diffusion alone is very inefficient to mix and disperse usual species: for instance, molecular diffusivity of carbon dioxyde in air is $16 \cdot 10^{-6}\,\mathrm{m}^2 \cdot \mathrm{s}^{-1}$, meaning that molecules separate at a rate of only a few millimeters per second.

A usual way to enhance mixing and dispersion consists in stirring the fluid in order to generate large scale uncorrelated turbulent structures, which act in a similar way (i.e. normally diffusive) as molecular diffusion, but with an enhanced diffusion coefficient $K_{\mathrm{turb}} \propto L\sigma$ with L the turbulence correlation length scale and σ the turbulent fluctuating velocity (standard deviation of the turbulent velocity field). In atmospheric dispersion for instance, the turbulent correlation length is typically of the order of hundreds of meters (let us take 100 m as an order of magnitude) with velocity fluctuations typically of the order of meters per second in normal conditions (let us take 1 m/s as an order of magnitude), leading to a turbulent diffusivity coefficient K_{turb} of the order of $30\,\mathrm{m}^2 \cdot \mathrm{s}^{-1}$, meaning that fluid particles separate at a rate of several meters per second, hence many orders of magnitude larger than molecular diffusion. The efficiency of turbulent diffusion therefore relies on the capacity of a substance to spread thanks to the uncorrelated motion of large scale turbulent eddies. However, if we consider the dispersion of a patch initially much smaller than the turbulent correlation scale L (for instance a patch with an initial dimension within the inertial range of the carrier turbulence, hence much smaller than the energy injection scale L, and larger than the dissipation scale η), another mechanism is necessary to allow the patch to grow first at sufficiently large scales to eventually undergo the effect of uncorrelated turbulent diffusion. Such an inertial scale mechanism is ensured by the *super-diffusive* nature of turbulence at inertial scales. Processes where the mean square separation grows faster than in normal diffusion (i.e. $< \vec{D}^2 > \propto t^\alpha$, with $\alpha > 1$) are called *super-diffusive*. Unlike normal diffusion, *super-diffusion* is generally associated with an heterogeneous and a non-gaussian growth of the spreading patch (Fig. 1).

The super-diffusive nature of turbulence was first emphasized by Lewis Richardson in his seminal 1926 article (see Richardson 1926). Richardson already noted in 1926 that "*a small dense cluster of marked molecules, represented by the dot in Fig. 2(a1) which, by molecular diffusion alone, would spread through the successive spherical clusters shown in Fig. 2(a2) and (a3), actually seldom passes through the large spherical stage Fig. 2(a3), because it is first sheared into two detached clusters as suggested in Fig. 2(a4). These are carried far from one another, and are likely to be again torn into smaller pieces as in Fig. 2(a5)*". This qualitative description by Richardson shows how turbulence acts to super-diffusively separate particles initially packed in a small patch, in order to create sufficiently large separations where the uncorrelated motion of turbulent eddies eventually disperses particles at large scales.

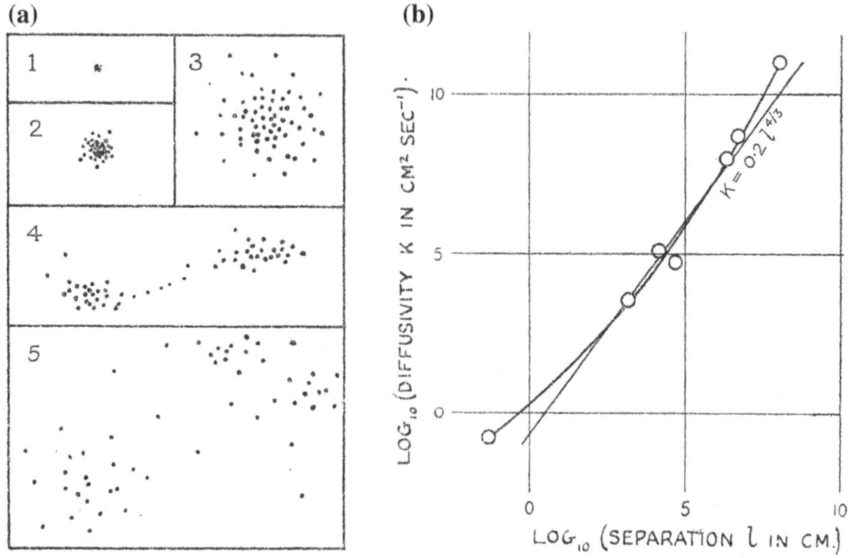

Fig. 2 **a** Qualitative illustration of the non-normal dispersion of a dense cluster of particles as proposed in Richardson's original 1926 article. **b** Original empirical derivation of the "4/3rd" law by Richardson. (Both figures are taken from Richardson's seminal article on relative dispersion Richardson 1926)

Richardson gave an interpretation of turbulent super-diffusion in terms of a non-Fickian process which could be locally modeled as a normal diffusion process, but with a scale dependent diffusion coefficient which depends on particles separation D, according to the celebrated Richardson's 4/3rd law: $K(D) \propto D^{4/3}$. Richardson conjectured such a scale dependent diffusive scenario from an empirical short time, scale by scale, analysis of local diffusion properties over a wide range of phenomena, from diffusion of oxygen intro nitrogen, to the diffusion of cyclones in the atmosphere (Richardson 1926) (Fig. 2b), such that at each scale the mean square separation could be locally written as $D^2 \propto K(D)t$. It is now accepted that his derivation of the 4/3rd law was at the same time fortuitous and the result of his unique intuition (Sawford 2001). Richardson also showed that such a non-Fickian diffusion resulted in a cubic super-diffusive growth of the mean square separation of pairs of particles according to the law $\langle D^2 \rangle = g\epsilon t^3$, where ϵ is the turbulent energy dissipation rate and g a universal constant since known as the Richardson constant.

2.2 *Batchelor and Richardson Regimes for Pair Dispersion*

Richardson's work was later refined by Batchelor and Obukhov in the 1950s (Batchelor 1950) thanks to Kolmogorov's 1941 phenomenology of turbulence. In K41 phenomenology, phenomena happening at within inertial scales of turbulence (i.e. scales much larger than dissipative scales (so that they are not affected by viscosity) and much smaller than the energy injection scale of the flow (so that they are not affected by large scale inhomogeneities and anisotropy) are statistically controlled by one single physical parameter, namely the mean energy dissipation rate ϵ. This framework then imposes simple dimensional constraints, which have been applied by Batchelor and Obukhov to the pair dispersion problem to show that the rate of separation of pairs of fluid elements in turbulent flows with initial separation $\vec{D}_0$ at inertial scales ($\eta \ll D_0 \ll L$) must obey the following scalings:

$$R^2 = \left\langle \left(\vec{D} - \vec{D}_0\right)^2 \right\rangle = \begin{cases} S_2(\vec{D}_0)t^2 & \text{if } t < t_0 \quad (3a) \\ g\epsilon t^3. & \text{if } t > t_0 \quad (3b) \end{cases}$$

with $S_2(\vec{D}_0)$ and t_0 both depend on the particles motion at scale D_0. In this scenario, the mean square separation of a pair of particles first grows ballistically (Eq. 3a) before transiting toward a faster t^3 regime (Eq. 3b) as predicted by Richardson. In the early ballistic regime (also known as the Batchelor regime), the separation rate depends on the initial separation $\vec{D}_0$ via the second order Eulerian structure function $S_2(\vec{r}) = \langle |\delta_{\vec{r}}\vec{u}|^2 \rangle$ estimated at the initial separation D_0 (with $\delta_{\vec{r}}\vec{u}$ the increment between two points separated by a vector $\vec{r}$ of the eulerian velocity field of the flow; note that homogeneity is assumed, so that velocity increment only depends on the separation vector). Formally speaking, the ballistic regime (Eq. 3a) is nothing but the leading term of the Taylor expansion for the mean square pair separation at short times, expressed in terms of the initial mean square relative velocity between particles (Batchelor 1950; Ouellette et al. 2006). Note that such a ballistic Taylor expansion is a general and purely kinematic relation valid for any early dispersion process and is not limited to the case of turbulence. Specificities of turbulence appear when expliciting the form of the structure function S_2 at inertial scales, which in K41 phenomenology is simply given (from dimensional considerations) by $S_2(r) \propto (\epsilon r)^{2/3}$. This ballistic regime is expected to hold only for times shorter than a characteristic time t_0 related to the persistence of the initial relative velocity, which for inertial scales of turbulence is related to the eddy turn-over time at scale D_0, $t_0 \propto r^{2/3}/\epsilon^{1/3}$. Qualitatively speaking the ballistic Batchelor regime can be seen as the result of the correlated motion of particles evolving in a same initial turbulent eddy with characteristic size D_0.

For times exceeding t_0, a transition is expected towards an enhanced dispersion regime, as predicted by Richardson. The mean square separation then grows as the cube of time. An important specificity of the Richardson regime is that the growth rate of pair separation does not depend anymore on the initial distance D_0 between the particles. The loss of memory of initial conditions is a reasonable assumption for long term diffusion. Under this assumption, the t^3 dependency can be understood in the framework of Kolmogorov 1941 phenomenology of turbulence (Kolmogorov 1941) (hereafter referred as K41) as a simple dimensional constraint. K41 states indeed that for sufficiently large Reynolds number, the only relevant physical parameter for the dynamics of turbulence at inertial scales is the average energy dissipation rate per unit mass ϵ (with dimensions [m^2 s^{-3}]): $D^2 \propto \epsilon t^3$ is then the only dimensionally consistent relation if initial separation is forgotten.

The Richardson constant g in Eq. 3b is one of the most fundamental constants in turbulence (together with the Kolmogorov constant C_K). It plays a major role in turbulent dispersion and mixing processes. However, in spite of its importance, it is only recently that estimations of its value started to converge towards a well accepted value (Sawford 2001; Salazar and Collins 2009). This is due to the difficulty to observe experimentally Richardson's superdiffusion (Sawford 2001; Bourgoin et al. 2006). Until recently, best estimates for g still spanned several orders of magnitude. Most recent high resolution direct numerical simulations seem to point toward an estimate of $g \sim 0.5$–0.6 (Bitane et al. 2012; Boffetta and Sokolov 2002) (Fig. 3a). This value is also consistent with experiments by Ott and Mann (2000) in homogeneous and isotropic turbulence at a moderate Reynolds number $R_\lambda \simeq 100$, based on Taylor micro-scale (Fig. 3b). However the procedure to extract the Richardson constant in these experiments has been controversial, as the authors had to introduce an arbitrary virtual time origin T_{origin} to find a reasonable cubic fit for the growth rate of the mean square separation $< (\vec{D} - \vec{D}_0)^2 > \propto \epsilon(t - T_{origin})^3$. As we will see later, such a virtual time origin can actually be justified on robust theoretical grounds and its value shown to be also related to the Richardson constant g. Apart from this experiment, high resolution data in well controlled laboratory experiments is still very scarce. This is very likely due to the difficulty in accessing sufficiently long tracks in Lagrangian measurements, allowing to unambiguously observer the long-term cubic regime *a la* Richardson. Recent experiments by Bourgoin et al. (2006)

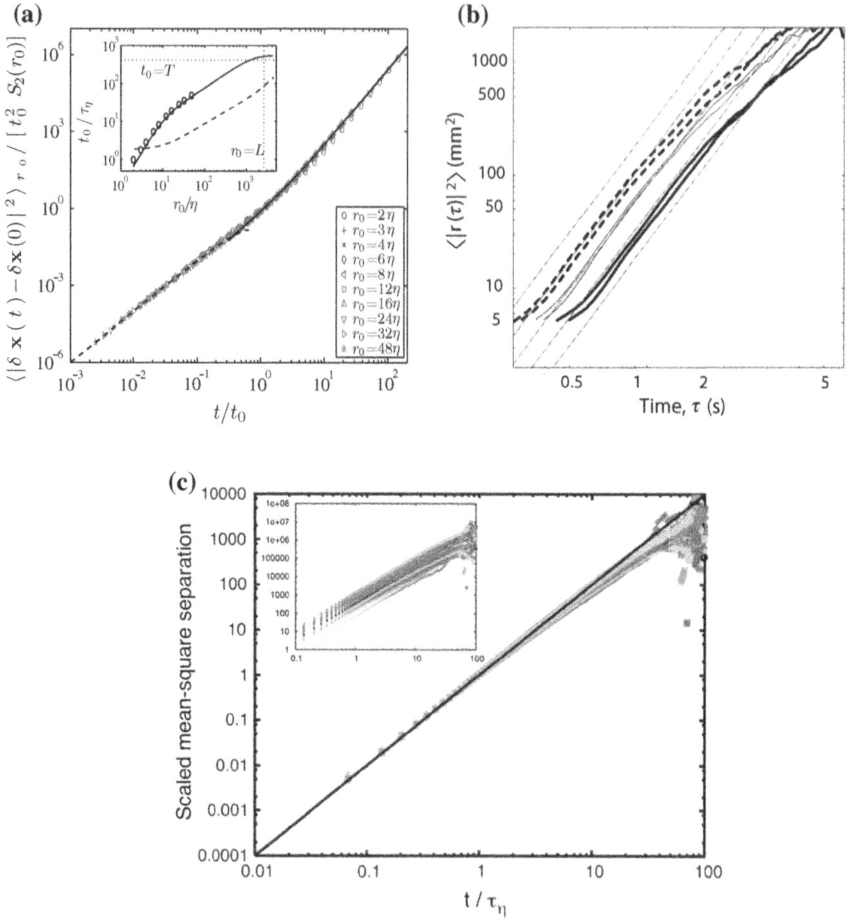

Fig. 3 Growth of the mean square separation of particle pairs in: **a** direct numericam simulations by Bitane et al. (2012) (time is non-dimensionalized by $t_0 = S_2(D_0)/2\epsilon$ which within K41 framework is proportional to $\epsilon^{-1/3}D_0^{2/3}$, different symbols correspond to different initial separations); at short time the separation follows a ballistic regime *a la* Batchelor, while at long times the separation is cubic, *a la* Richardson, with a transition occurring for times $t \simeq t_0$. **b** Experiments by Ott and Mann (2000); time in this plot has been shifted by virtual time origin T_0 to emphasize the cubic behavior. **c** experiments by Bourgoin et al. (2006) where the ballistic regime is robustly observed, without any adjustable parameter

report highly resolved particle tracking measurements of relative dispersion at high Reynolds numbers (up to $R_\lambda \simeq 800$). However only the early ballistic regime 3a was observed, and no hint toward a cubic regime was found (Fig. 3c).

As already mentioned, in his seminal 1926 article (Richardson 1926), Richardson empirically related such a superdiffusive regime to a non-Fickian process, with a local diffusivity coefficient K which depends on the spatial scale D: $K(D) \propto D^{4/3}$.

Further refinements in the framework of K41 phenomenology extended Richardson's non-Fickian phenomenology by considering also a time scale dependency of the local diffusivity coefficient (Boffetta and Celani 2000; Klafter et al. 1987; Salazar and Collins 2009) such that $K(D,\tau) = k_0 \epsilon^\gamma D^\alpha \tau^\beta$ (with the dimensional constraints that $3\alpha + 2\beta = 4$ and $\gamma = 1 - \alpha/2$). Such processes also lead to a t^3 regime for the mean square separation (the Richardson constant g is then directly related to k_0). However no clear physical interpretation for the origin of such a time/scale dependency of the local diffusivity is still known.

In the next subsection, a simple phenomenology for pair dispersion is proposed which elucidates many aspects of the problem, in particular regarding the origin of the cubic super-diffusive long-term regime and its deep connection with fundamental properties of the turbulent energy cascade (Bourgoin 2015).

2.3 A Simple Ballistic Phenomenology of Turbulent Superdiffusion

Let consider an ensemble of particles with an initial mean square separation D_0^2. In the short term, the particles will separate ballistically according to the kinematic relation (3a). Let t_0 be the typical duration of the ballistic initial ballistic growth. Experimental and numerical evidence discussed in the previous section confirme Batchelor's suggestion that $t_0 \propto \epsilon^{-1/3} D_0^{2/3}$. Following Bitane et al. we will write $t_0' = \alpha S_2(D_0)/2\epsilon$, with α a non-dimensional parameter we shall call *persistence parameter* as it quantifies the persistence of the ballistic regime. The main idea behind the dispersive process proposed here is that of a ballistic cascade mechanism, as illustrated in Fig. 4. It is based on the trivial idea that if an ensemble of particles with initial mean square separation $\vec{D}_0^2$ starts to disperse ballistically, with a separation rate $S_2(\vec{D}_0)$ over a given period t_0' after which it reaches a new mean square separation $\vec{D}_1^{\;2}$, instead of considering a sudden transition toward an enhanced cubic dispersion regime *à la* Richardson (as in Eq. 3b), the ballistic process can instead be iterated starting from the new mean square separation $\vec{D}_1^2$, with a new separation rate $S_2(\vec{D}_1)$ over a period t_1' and so on. Thus, in this scenario the time evolution of particles mean square separation is simply described by the iterative process:

$$D_{k+1}^2 = D_k^2 + S_2(D_k) t_k^2(D_k), \tag{4}$$

where $D_k^2 =< |\vec{D}_k|^2 >$ represents the mean square separation of pairs after the kth iteration step, $t_k'(D_k)$ is a scale dependent "time of flight" characteristic of the duration of the ballistic motion at step $k+1$. For the case of turbulent flows, $S_2(D_k)$ and $t_k'(D_k)$ will be prescribed later by imposing K41 scalings.

A concrete implementation of the iterative scheme (4) requires the expressions for the scale dependent separation rate $S_2(D_k)$ and the ballistic time of flight $t_k'(D_k)$ tobe

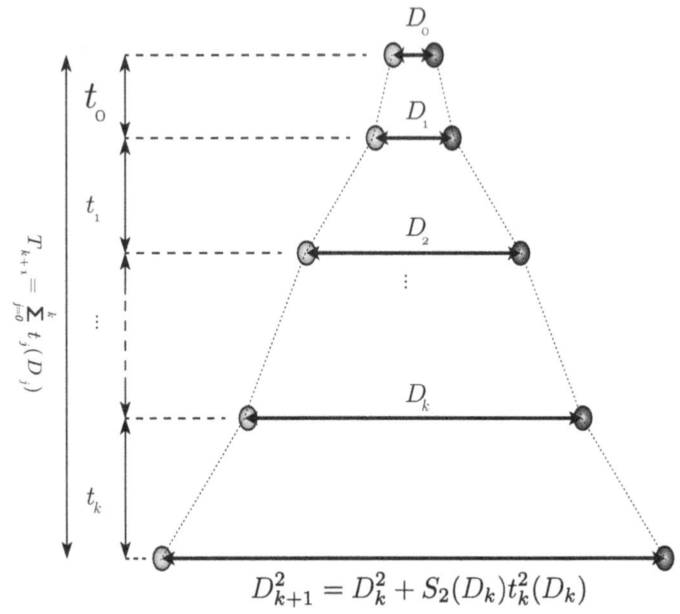

Fig. 4 Illustration of the iterative ballistic cascade for the relative separation of particles with initial mean square separation D_0^2: at each iteration step k, the mean square separation between particles grows ballistically from D_k^2 to D_{k+1}^2 with a growth rate $\sqrt{S_2(D_k)}$, during a time lag t'_k. The overall time required to reach separation D_k at the iteration number k is $T_k = \sum_{j=0}^{k-1} t'_j(D_j)$

specified. For particles with separation in the inertial range of scales of the carrier turbulence, under the local isotropy assumption, the structure function is known to follow the K41 scaling:

$$S_2(\vec{D}_0) = \frac{11}{3} C_2 \epsilon^{2/3} D_0^{2/3}, \tag{5}$$

with $D_0 = |\vec{D}_0|$ and where C_2 is a universal constant with a well-known value of approximately 2.1; as C_2 is analytically related to the Kolmogorov constant $C_K \simeq C_2/4$ (Sreenivasan 1995) characterizing the celebrated $-5/3$ spectrum of turbulent kinetic energy ($E(k) = C_K \epsilon^{2/3} k^{-5/3}$), we shall refer to $C = \frac{11}{3} C_2 \simeq 7.7$ as *the* Kolmogorov constant in the sequel.

The iterative ballistic process can now be explicitly written as

$$D_{k+1}^2 = D_k^2 + S_2(D_k) t_k'^2(D_k) \quad \text{with} \quad \begin{cases} S_2(D_k) = C\epsilon^{2/3} D_k^{2/3} \\ t'_k(D_k) = \alpha t_k = \alpha S_2(D_k)/2\epsilon \end{cases}, \tag{6}$$

where the Kolmogorov constant C and the persistence parameter α are the only parameters. The Kolmogorov constant being known, α is the only adjustable parameter of the problem. The iterative process Eq. 6 leads to a simple geometrical progression where

$$\begin{cases} D_{k+1}^2 = AD_k^2 \\ t'_{k+1} = A^{1/3}t'_k \end{cases}, \tag{7}$$

with

$$A = 1 + \frac{\alpha^2 C^3}{4}. \tag{8}$$

Simple arithmetics then lead to the following relations for the growth of the mean square separation as a function of the total time $T_k = \sum_{j=0}^{k-1} t'_j(D_j)$:

$$D_k^2 = g\epsilon \left[T_k + \left(\frac{D_0^2}{g\epsilon} \right)^{1/3} \right]^3 \tag{9}$$

with

$$g = \left[2\frac{A^{1/3} - 1}{\alpha C} \right]^3 = \left[2\frac{(1 + \frac{\alpha^2 C^3}{4})^{1/3} - 1}{\alpha C} \right]^3 \tag{10}$$

Several points are worth being noted at this point:

- Relation Eq. 9 shows that in the long term ($T \gg \left(D_0^2/g\epsilon\right)^{1/3}$), the iteration of elementary K41 scale dependent ballistic steps eventually builds a Richardson cubic regime where $D^2 = g\epsilon T^3$.
- It also shows that at intermediate times, the growth follows a similar cubic regime, but with a negative virtual time origin $T_{origin} = -\left(D_0^2/g\epsilon\right)^{1/3}$ such that $D^2 = g\epsilon(T - T_{origin})^3$, what supports the empirical approach by Ott and Mann (2000) to extract the value of the Richardson constant from short experimental tracks.
- Finally, relation Eq. 10 related the Richardson constant to the Kolmogorov constant C and to the persistence parameters.

Figure 5 shows the result of the ballistic phenomenology for the mean square separation as a function of time as given by Eq. 9, compared to the direct numerical simulations by Bitane et al. (2012). The well accepted value $C = 11C_2/3$ with $C_2 \simeq 2.1$ has been used for the Kolmogorov constant, while the persistent parameter has been adjusted in order to retrieve the well accepted value $g \simeq 0.55$ for the Richardson constant, based on Eq. 10, what leads to an optimal value $\alpha \simeq 0.12$. It can be seen that the iterative ballistic phenomenology reproduces almost perfectly the numerical data. The first interesting observation concerns the collapse (as already noted by Bitane et al. 2012) of the numerical data for all initial separations within inertial scales, which is perfectly follows the prediction of the ballistic phenomenology, both for the short term ballistic regime and the long term cubic regime. Besides, not only the global trend of the mean square separation evolution is very well described by the model (both for short and long term regimes), but some subtler details are also well captured. For instance, as in the simulation, the transition between the early Batchelor regime and the Richardson regime is robustly found to occur around

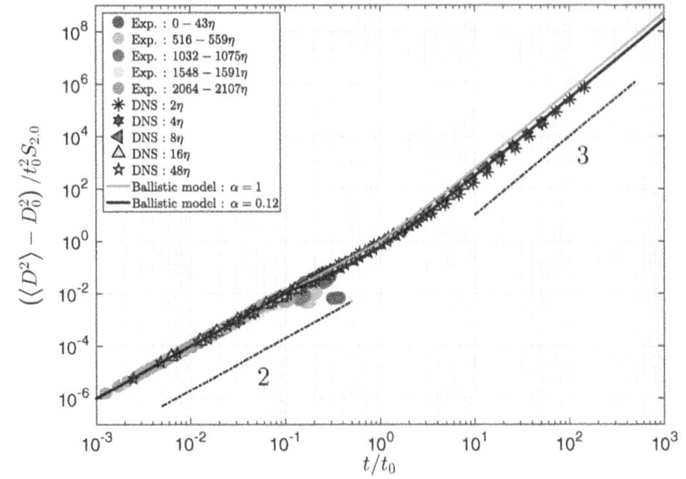

Fig. 5 Comparison of the prediction from the ballistic model for two values of the persistence parameter ($\alpha = 1$ and $\alpha = 0.12$) and for the mean square separation of particles with different initial separation D_0 (given in the inset) within inertial scales, obtained in direct numerical simulations (Bitane et al. 2012) and in high resolution particle tracking experiments (Bourgoin et al. 2006). Note that the time axis is normalize by $t_0 = S_2(D_0)/2\epsilon$ and the mean square separation is normalized by $t_0^2 S_2(D_0)$

$t = t_0$, even if the duration of the initial ballistic iteration is $t_0' = \alpha t_0 \ll t_0$ (we recall that $\alpha \simeq 0.12$ in the present implementation of the ballistic model). A mild slowing down of the ballistic separation for $t \lesssim t_0$ (prior to the transition towards the cubic regime), present in the simulations is also captured by the iterative ballistic model.

Figure 5 also shows the mean square separation of pairs measured in high resolution particle tracking experiments by Bourgoin et al. (2006). In those experiments, only the Batchelor ballistic regime was reported, while no hint of Richardson regime was detected. Figure 5 emphasizes a possible reason for the failure in experiments to observe the Richardson regime: the longest experimental tracks did not exceed a few tenth of t_0 while the separation needs to be tracked for at least a few t_0 to reasonably detect the transition toward the cubic regime. A simple possible strategy to improve the chances to observe the cubic regime in experiments would simply consist in better controlling the injection of particle pairs in order to achieve sufficiently small initial separations, hence reducing the time t_0 required for the transition to occur within experimentally accessible tracking time.

To conclude this first part on relative dispersion of Lagrangian tracers, it should be mentioned that the present phenomenology can be pushed event further and can be extended to account for time asymmetry of relative dispersion (Sawford et al. 2005; Jucha et al. 2014; Berg et al. 2006) as well as known differences between 3D and 2D turbulence (Faber and Vassilicos 2009). These aspects go beyond the scope of the present chapter, but the interested reader can refer to the recent article (Bourgoin 2015) for more details on these extensions.

3 Preferential Concentration of Inertial Particles in Turbulence

Previous chapter was dedicate to the dynamics of fluid tracers, which are typically neutrally buoyant particles significantly smaller than the dissipation kolmogorov scale of the flow η. In the present chapter the collective motion of material particles which do not necessarily fulfill these conditions is discussed. Such particles, with density different than that of the carrier fluid and/or size larger than η are expected to behave differently than fluid tracers and are generically called *inertial particles*. A striking feature observed in such situation, is the trend to preferential concentration, which has been reported for long (Squires and Eaton 1991; Fessler et al. 1994) and which is still thoroughly studied (Saw et al. 2008; Salazar et al. 2008; Scott et al. 2009; Olla 2010; Bourgoin and Xu 2014). The present chapter will focus on diagnosis of preferential concentration phenomenon and on the main known properties of clustering of inertial particles in turbulence. Possible mechanisms for the origin of preferential concentration as well as possible impacts on the collective dynamics of the particles (in particular regarding collective settling) will be briefly discussed.

3.1 Diagnostics of Preferential Concentration with Voronoï Tessellation

Until recently most usual approaches to quantify the level of clustering are based on the pair correlation function while box counting methods are preferred to access local concentration fields. We will focus here on a method, based on Voronoï tessellation, which has been recently introduced to the field and which revealed to be particularly well suite to investigate the preferential concentration phenomenon. A full and detailed comparison of different methods can be found in the review by Monchaux et al. (2012).

A Voronoï diagram is the unique decomposition of 2D space (the concept can be generalized to N-dimensional space) into independent cells associated to each individual particle. One Voronoï cell is defined as the ensemble of points that are closer to a particle than to any other. Use of Voronoï diagrams is very classical to study granular systems and has also been used to identify galaxies clusters. Voronoï diagrams computation is very efficient with the typical number of particles per image (up to several thousands) we have to process. Figure 6a, b show a raw acquired image as well as the located particles and the associated Voronoï diagram.

Why using Voronoï tessellations? From the definition of the Voronoï diagrams, it appears that the area A of a Voronoï cell is the inverse of the local 2D-concentration of particles; therefore the investigation of Voronoï areas field is strictly equivalent to that of local concentration field. Let me recall that usually local concentration fields are obtained through box counting methods (Aliseda et al.

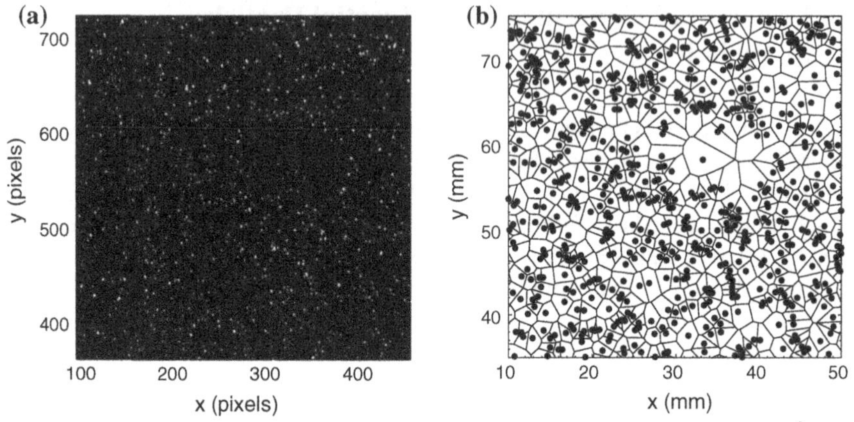

Fig. 6 **a** A typical raw image. **b** Particles located in this image and the associated Voronoï diagram. For clarity, we show only one third of the full acquired image

2002) which show several disadvantages: they are computationally inefficient and they require to select an arbitrary length scale (the box size), whereas in Voronoï diagrams computation, no length scale is a priori chosen and the resulting local concentration field is obtained at an intrinsic resolution. Similarly, the pair correlation function only gives global (non local) information and is also associated to the choice of a length scale that spans the whole values of interest increasing dreadfully the computation time. Finally, another interest of Voronoï diagrams is that as each individual cell is associated to a given particle at each time step, thus tracking in a Lagrangian frame the particles directly gives access to the Lagrangian dynamics of the concentration field itself along particles trajectories. Though such Lagrangian aspects will not be discussed here, they represent an important opening which shall be addressed in future studies in the field.

Some relevant properties of Voronoï diagrams. Whatever the measurement and data analysis technique used, when one refers to preferential concentration, it is implicitly assumed that one deals with statistical preferential concentration compared to the case where particles would be spatially distributed as a random Poisson process (RPP). In order to quantify preferential concentration, the Probability Density Function (PDF) of the measured Voronoï areas for each experiment is compared to that expected for a RPP. Main known properties of Voronoï diagrams associated to RPP can be found in a short review by Ferenc and collaborators (2007) and references herein. The first moment of Voronoï areas PDF has nothing to do with the spatial organization of the particles since the average Voronoï area, $\overline{A}$, is always identical to the mean particles concentration. Therefore, in the sequel I will generally focus on the distribution of the normalized Voronoï area $V = A/\overline{A}$ which is of unit mean. The only known exact result for RPP Voronoï areas statistics concerns the second order moment in the 2D case which is equal to $< V^2 >_{\mathrm{RPP}} = 1.280$, corresponding to a standard deviation $\sigma_V^{\mathrm{RPP}} = \sqrt{< V^2 >_{\mathrm{RPP}} - 1} \simeq 0.53$. Regarding the shape of the

PDF of Voronoï areas statistics for a RPP, no analytical solution is known (most of the authors fit them with Gamma distributions). Ferenc and collaborators propose a compact analytical expression involving the space dimension as a single parameter: this analytical expression is used here as a RPP reference.

3.2 *Main Properties of Preferential Concentration in Turbulence*

In this section, the use of Voronoï tessellation and the main properties of particles clustering in turbulence is described using as reference the experimental data in Monchaux et al. (2010), Obligado et al. (2014), obtained for water droplets in wind tunnel with passive and active grid generated turbulence. Reynolds number in those experiments span a broad range $R\lambda \in [70, 400]$ while particles Stokes number spans the range $St \in [0.2, 10]$.

Experimental Voronoï areas distributions. Figure 7a displays the PDFs of the dimensional Voronoï cells area, A, for 40 different experiments. When the dimensional area A is considered, one observes that the maximum of these PDFs spans over two decades. This evolution is representative of the average number of particles per image (or equivalently of the global seeding concentration C_0) which for the ensemble of experiments represented goes from 50 to 5000. Note that as the average number of particles per image decreases (i.e. as the mean Voronoï area increases), the scatter of the right tail on these PDFs increases as a consequence of the lesser statistical convergence. As shown in Fig. 7b, all these PDFs collapse reasonably if we consider the centered-reduced PDF of the logarithm of normalized Voronoï areas. This indicates that while Voronoï area PDFs of RPP are usually described by Gamma distributions (Ferenc and Neda 2007) they are well described by a log-normal distribution for the investigated inertial. As seen on the figure, superimposition with a log-normal distribution is almost perfect in the interval $\pm 2\sigma_{\log(V)}$, where $\sigma_{\log(V)}$ stands for the standard deviation of $\log(V)$ (note that extreme PDF tails for statistics of the logarithm of V are beyond experimentally accessible statistical convergence). To date we do not have any theoretical interpretation of this log-normality, but this result shows that in the limit of experimental convergence, normalized Voronoï areas PDFs can be described with one single scalar quantity which we choose here to be the standard deviation of the normalized Voronoï areas σ_V.

Quantifying preferential concentration. In Fig. 7c we present the evolution of normalized Voronoï areas standard deviation σ_V as a function of Stokes number in the wind tunnel experiments. As mentioned earlier, the standard deviation of Voronoï areas for a RPP is analytically known to be $\sigma_V^{\mathrm{RPP}} \simeq 0.53$ which defines a reference value to compare with. A standard deviation σ_V significantly exceeding 0.53, reveals the existence of high and low concentration events compared to the RPP case. Oppositely, a standard deviation σ_V below this reference value would evidence the tendency of particles to distribute in a more organized arrangement ($\sigma_V = 0$ in

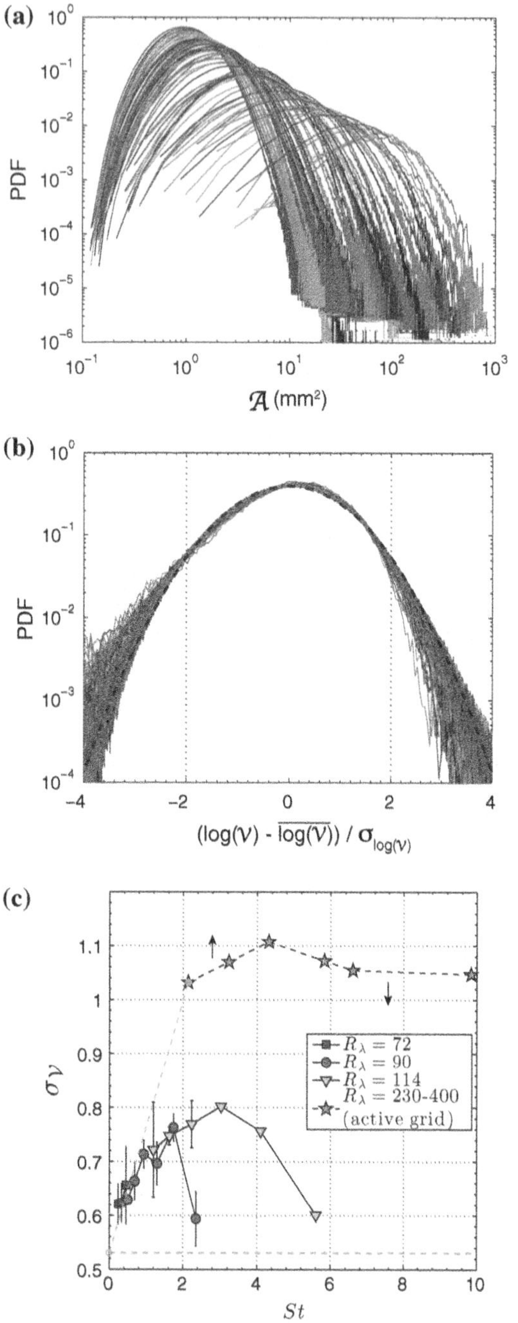

Fig. 7 **a** PDF of dimensional Voronoï area A for 40 experiments spanning all R_λ, St and volume loading explored. **b** Centered and normalized PDF of the logarithm of Voronoï area for the 40 experiments from upper figure; *black dashed line* represents a gaussian distribution. **c** Standard deviation of Voronoï areas as a function of average Stokes number. Lines connect different experiments for which Reynolds number (Error-bars represent the dispersion between experiments with different C_0) (color figure online)

the limit of a perfect crystal). As seen on Fig. 7c, for the range of explored Stokes numbers (spanning from 0.2 to almost 10) the standard deviation of the normalized Voronoï areas always exceeds 0.57 and reaches values as high as 1.1, what shows that preferential concentration is always present over this range of Stokes numbers, consistently with former experimental and numerical investigations. Furthermore, the curves in Fig. 7c also show a mild peak for values of the Stokes number in the range $St_{\text{pk}} \simeq 2 - 4$ with a possible dependency of the peak position on Reynolds number. However existing data does not allow to draw definitive conclusions regarding the Reynolds number dependency, in particular for the case of large values of Reynolds number (obtained in active grid experiments). This is due to the fact that in these experiments, the Reynolds and Stokes number could not be varied independently. Robust results concern however the maximum of clustering for Stokes number of order unity (in agreement with most existing studies (Monchaux et al. 2012) as well as the enhancement of clustering with increasing Reynolds number.

Clusters definition and identification. An important issue when it comes to investigate and characterize clustering properties of particles concerns the definition of *what a cluster is*. Usual approaches consist in prescribing a threshold in local concentration C_c and defining clusters as patches as the connected regions where the concentration exceeds this threshold. Voronoï tesselations are particularly well suited to achieve this. Voronoï areas PDF may indeed be used to identify clusters of particles as follows. Consider Fig. 8a which presents the Voronoï PDF for a typical experiment and for a RPP. These PDFs intersect twice (which is more visible on Fig. 8b showing the ratio of both PDFs): for low and high values of normalized Voronoï area, corresponding respectively to high and low values of the local concentration, experimental PDF is above the RPP one, while we observe the opposite for intermediate area values. This is consistent with the intuitive image of preferential concentration: inertial particles concentration field is more intermittent than the RPP, with more probable *preferred* regions where concentration is higher than the Poisson case and subsequently also more probable *depleted* regions where concentration is lower than in the Poisson case. These intersection points V_c and V_v can be taken as an intrinsic definition of particle *clusters* and *voids*: for a given experiment, Voronoï cells whose area is smaller than the first intersection V_c are considered to belong to a cluster while those whose area is larger than the second intersection V_v are associated to voids. I insist on the fact that these thresholds are intrinsically chosen experiment wise and so vary from one experiment to another; in particular their value depends on the number of particles per image and their evolution with the seeding concentration C_0 is find to be affine. Figure 8c displays a full Voronoï diagram corresponding to one image taken from one experiment. On this diagram, cells corresponding to clusters (*resp.* to voids) have been colored in dark gray (*resp.* light gray) while the remaining cells have been patched with white. It appears that dark gray cells (*resp.* light gray cells) tend to be connected in groups of various sizes and shapes that are identifed as clusters (*resp.* voids) whenever they belong to the same connected component.

Clusters geometry. Figure 9a shows the PDFs of the area A_C of clusters, normalized by the average area $\langle A_C \rangle$ (clusters and voids present similar geometric characteristics,

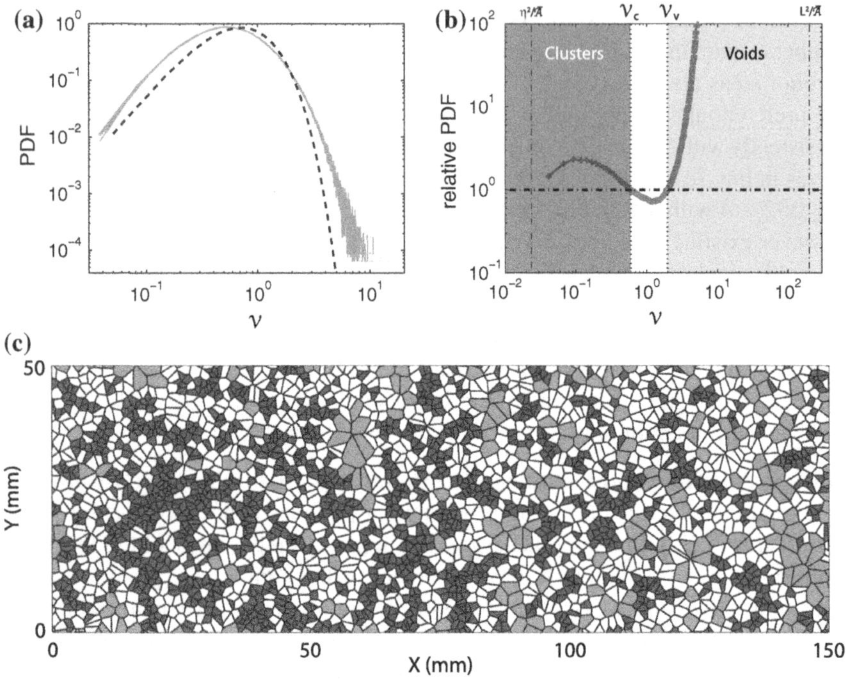

Fig. 8 A way to identify clusters: **a** Superposition of the Voronoï areas PDF for a typical experiment ($R_\lambda = 85$, $St = 0.33$, $C_0 = 500$ particles per image); 10 continuous lines associated to ten sets of 500 UIVD are represented (dispersion is negligible) and a RPP (*dotted line*); **b** ratio of the two PDF presented on the *left figure*. Vertical *dash-dotted lines* indicates η^2 (*left*) and L^2 (*right*). **c** colored visualization of clusters (*dark grey*) and voids (*light gray*) (color figure online)

hence only results for the clusters will be shown). Interestingly, all PDFs collapse onto a single curve. For large areas the PDFs of clusters area follow a power law with an exponent close to -2 (Monchaux et al. 2010; Obligado et al. 2014). These PDFs exhibit a clear maximum, with a peak around a most probable area $A_C^{max} \simeq 0.2\langle A_C\rangle$, evidencing the fact that clusters do have a *typical characteristic area*. Figure 9b shows the typical linear dimension $\sqrt{\langle A_C\rangle}$ of the clusters normalized by the dissipation scale η of the carrier flow, as a function of Stokes number, showing that the typical cluster size is of the order of 10η. Finally, one important property concerns the fractal geometry of cluster. This is evidenced by computing the joint histogram, for all clusters of the clusters perimeter P_C and of the root square of its area $A_C^{1/2}$, as shown in Fig. 9c. For non fractal structures we would expect $P_C \propto A_C^{1/2}$ while here the main trend is $P_C \propto A_C^{\alpha/2}$, with the exponent $\alpha \gtrsim 2$ evidencing the fractal nature of clusters. Moreover, the almost continuum range for this exponent (ranging from $\sim$2 to $\sim$3.5) evidence again the extreme complexity of this structures.

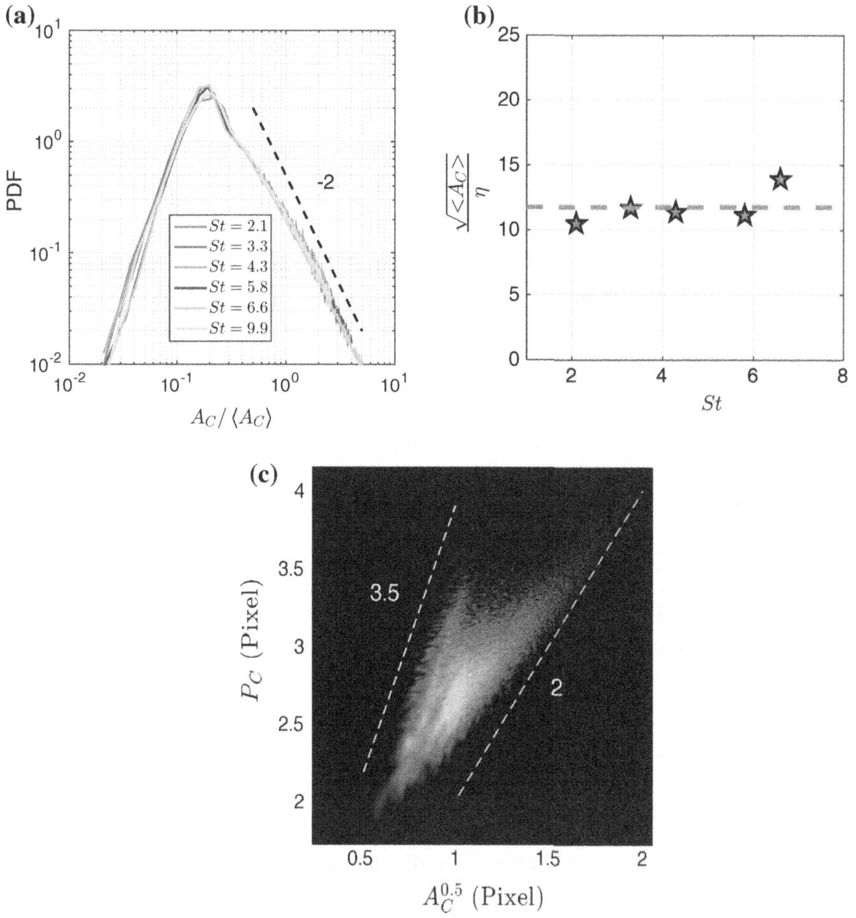

Fig. 9 **a** PDFs of clusters area normalized by the mean. **b** Mean linear dimension of clusters normalized by the dissipation scale of the flow as a function of Stokes number. **c** Joint histogram of clusters perimeter and the square root of their area obtained experimentally (axes are in logarithmic scales so that the emphasizes slopes correspond to power laws dependencies). Data in these figures are from Obligado et al. (2014)

3.3 *Origins of Preferential Concentration*

The origin of preferential concentration of inertial particles in turbulence can be understood at several levels, from very fundamental and generic consideration related to the dissipative nature of particle dynamics to intuitive interactions between the particles and the structures of the carrier turbulence. The following paragraphs briefly address some of these aspects.

Inertial particles in turbulence as a dissipative dynamical system. Let consider a system whose state is described by a point $\vec{q} = (q_1, q_2, \dots)$ in a given state space Ω. For a *dynamical system* the state can evolve in time and the evolution from one state to another follows a vectorial differential equation given by a *rule* f such that:

$$\frac{\mathrm{d}\vec{q}}{\mathrm{d}t} = \vec{f}(\vec{q}) \tag{11}$$

with $\vec{q} \in \Omega$ and with the initial condition $\vec{q}(t=0) = \vec{q}_0$. If now we consider an ensemble of initial conditions, defined by an initial probability density of states $\rho(\vec{q}, t=0)$, so that $\int_\Omega \rho(\vec{q}, t=0)\mathrm{d}\vec{q} = 1$ and $\rho(\vec{q}, t=0)\mathrm{d}\vec{q}$ represents the probability to find at $t = 0$ the system in a elementary volume $\mathrm{d}\vec{q}$ around the state $\vec{q}$. A linear perturbation calculation then allows to derive the evolution equation for the density of states $\rho(\vec{q}, t)$ at $t > 0$:

$$\frac{\mathrm{D}\rho}{\mathrm{D}t} = \partial_t \rho + \vec{f} \cdot \vec{\nabla}\rho = -\rho\vec{\nabla} \cdot \vec{f} \tag{12}$$

which is nothing but a continuity equation ensuring that at all times $t > 0$, $\rho(\vec{q}, t)$ defines a probability density of states such that $\int_\Omega \rho(\vec{q}, t)\mathrm{d}\vec{q} = 1$. Note that $\frac{\mathrm{D}_{\vec{f}}\rho}{\mathrm{D}t}$ can be seen as the Lagrangian derivative in the flow $\vec{f}$. Two important cases are then to be distinguished depending on the value of $\vec{\nabla} \cdot \vec{f} = \partial_{q_i} f_i$:

- if $\vec{\nabla} \cdot \vec{f} = 0$, then Eq. 12 simply becomes $\frac{\mathrm{D}\rho}{\mathrm{D}t} = 0$, which is a continuity equation for the incompressible flow $\vec{f}$, so that volumes in parameter space are conserved. In particular, if initial states are uniformly distributed ($\rho(\vec{q}, 0)$ is constant and equal to $V_\Omega^{-1} = \left(\int_\Omega \mathrm{d}\vec{q}\right)^{-1}$), then the system remains uniformly distributed at all subsequent times. This result is known as the Liouville theorem. Dynamical systems such that $\vec{\nabla} \cdot \vec{f} = 0$ are called *conservative*.
- if $\vec{\nabla} \cdot \vec{f} < 0$, then $\frac{\mathrm{D}\rho}{\mathrm{D}t} > 0$ and the density along the Lagrangian trajectories in the state space eventually grows exponentially. As the integral of ρ over the whole state space is 1, that means that the density of state in other regions must decrease. In particular, if the initial density is uniform, this will result in the long term to regions of high density of states and depleted regions. Dynamical systems such that $\vec{\nabla} \cdot \vec{f} < 0$ are called *dissipative*. The resulting contraction of the state space is then related to the existence of *attractors* in state space towards which the system converges.

This simple generic mechanism can be applied to the dynamics of Lagrangian tracers and inertial particles in turbulence:

- for Lagrangian tracers, the dynamic system is simply defined by the state variable $\vec{q} = \vec{x}$, the position of the particles, the flow $\vec{f} = \vec{u}$ being simply the fluid flow itself, such that the dynamical system evolution equation (Eq. 11) simply writes $\dot{\vec{x}} = \vec{u}(\vec{x}, t)$. For incompressible flows, $\vec{\nabla} \cdot \vec{u} = 0$, so that in the dynamical system terminology the system is conservative (Eq. 12 is simply then the mass conservation equation): an initially uniform distribution of particles remains uniform. As expected Lagrangian tracers in incompressible flows do not form cluster.

- for inertial particles, on the contrary, the dynamical system is dissipative. This can be simply understood by considering the minimal Stokesian model given by Eq. 2, where particles dynamics is coupled to that of the fluid by Stokes drag. In that case, we can define the dynamical system based on the state variable $\vec{q} = (q_1 = \vec{x}, q_2 = \vec{v})$ in position-velocity state space, the flow function being $\vec{f} = (f_1 = \vec{v}, f_2 = \tau_p^{-1}(\vec{u} - \vec{v}))$. It is then trivial to show that $\vec{\nabla} \cdot \vec{f} = \partial_{q_1} f_1 + \partial_{q_2} f_2 = -d(\tau_p)^{-1} < 0$ (with the dimension of physical space). As a consequence the system is dissipative, according to dynamical space terminology, so that an initially uniform distribution of particles will tend to contract leading to the emergence of clusters (in position-velocity state space).

Therefore, the intimate dissiaptive nature of inertial particles dynamics already sustains a very fundamental mathematical reason for clustering to happen in turbulence (see for instance Mehlig and Wilkinson 2004; Mehlig et al. 2005; Bec et al. 2007 among others for more details). It does not tell though what the attractors in physical space are. These directly related to the topology of the carrier flow $\vec{u}$ and of the response time τ_p (which are the two parameters defining the flow function $\vec{f}$). The clustering properties are then naturally expected to depend on the structure of the carrier flow and the way it interacts with the particles.

The role of particle/turbulence interaction on the clustering of inertial particles clustering. The usual intuitive interpretation of clustering of inertial particles rely on the centrifugal expulsion of heavy particles from the core of turbulent eddies (light particles, as bubbles in water tend in the contrary to migrate towards the center of the eddies). As a consequence, dense particles are expected to cluster preferentially in low vorticity regions, while light particles are expected to explore preferentially high vorticity region (this last property being commonly used in experiment for instance to visualize vortex filaments by seeding a flow of liquid with small gas bubbles). This intuitive phenomenology has been formalized by Maxey (1987), in the limit of particles with small Stokes number ($St = \tau_p/\tau_\eta \ll 1$), where at first order particle's acceleration can locally by assumed to be comparable to fluid's acceleration. Then Eq. 2 implies that at first order in τ_p:

$$\vec{v} = \vec{u} - \tau_p \vec{a} \tag{13}$$

where the acceleration of the particles, $\mathrm{d}\vec{v}/\mathrm{d}t$, is locally assimilable to that of the fluid, $\vec{a} = \mathrm{d}\vec{u}/\mathrm{d}t$. Taking the divergence of this relation, it can be shown that

$$\vec{\nabla} \cdot \vec{v} = -\tau_p \vec{\nabla} \cdot \vec{a} \tag{14}$$

$$= -\tau_p \vec{\nabla} \cdot (\vec{u} \cdot \vec{\nabla} \vec{u}) \tag{15}$$

$$= -\tau_p (s_{ij} s_{ij} - 4 r_{ij} r_{ij}) \tag{16}$$

where $s_{ij} = \frac{1}{2}(\partial_j u_i + \partial_i u_j$ is the flow strain tensor and $r_{ij} = \frac{1}{2}(\partial_j u_i - \partial_i u_j$ the vorticity tensor. Note that in the last equation in Eq. 18, Einstein sommation convention is considered). This last equation shows that the velocity field $\vec{v}$ of inertial particles is compressible towards regions of high strain and low vorticity which particles can be expected to explore preferentially, as intuitively expected for dense inertial particles. The case of light particles requires to take into account the extra added mass term $\beta D\vec{u}/Dt$ in the equation of motion of the particles, with $\beta = 3\rho_f/(\rho_f + 2\rho_p)$. Equation 13 then becomes

$$\vec{v} = \vec{u} - \tau_p(1 - \beta)\vec{a} \tag{17}$$

and the divergence of $\vec{v}$ is then:

$$\vec{\nabla} \cdot \vec{v} = -(1 - \beta)\tau_p(s_{ij}s_{ij} - 4r_{ij}r_{ij}) \tag{18}$$

For dense particles ($\rho_p > \rho_f$) $\beta < 1$ so that as previously $\vec{\nabla} \cdot \vec{v} < 0$ in high strain/low vorticity regions. For light particles ($\rho_p < \rho_f$) $\beta < 1$ and $\vec{\nabla} \cdot \vec{v} < 0$ in low strain/high vorticity regions. Light particles then concentrate preferentially rotating regions of the flow.

Recently, Vassilicos and collaborators proposed an alternative mechanism (Goto and Vassilicos 2008; Coleman and Vassilicos 2009) to the usual *centrifugation* process just described. This mechanism relies on the simple idea, that Eq. 13 shows that in regions of the flow where the fluid acceleration vanishes ($\vec{a} = \vec{0}$)), namely *zero acceleration points*, the velocity of the particles is equal to that of the fluid. As in a first approximation the acceleration of the particles is also locally assimilable to that of the fluid, particles *stick* preferentially to zero acceleration points, with which they are swept in the flow (with locally a constant velocity, as acceleration is null). This process, called *sweep-stick* mechanism therefore predicts that particle preferentially concentrate in regions of the flow with low acceleration (rather than low vorticity), as shown in Fig. 10. Based on numerical simulations, the authors have shown that the centrifugation mechanism, where particles concentrate in low vorticity/high strain regions dominates for particles with small Stokes number, while sweep-stick dominates for particles with moderate Stokes number. This can be qualitatively understood, again from Eq. 13, considering that when the response time of the particles is sufficiently small $\vec{v} \simeq \vec{u}$ anywhere in the flow (not only near zero acceleration points), whereas for large values of the response time, the sticking condition $\vec{v} \simeq \vec{u}$ is only satisfied in regions where the acceleration is indeed small. The sticking specificity of zero acceleration points is therefore enhanced for highly inertial particles.

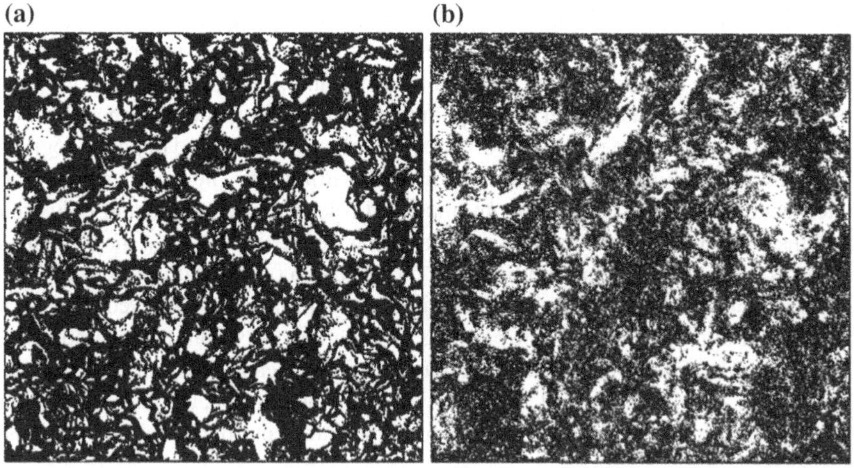

Fig. 10 Comparison of the position of inertial particles with $St = 2$ in numerical simulations (the stokesian minimal model Eq. 2 is used for the equation of motion of the particles) and the position of zero acceleration points of the carrier fluid (figure taken from Coleman and Vassilicos 2009)

4 Conclusion

The collective dynamics of particles in turbulence is a rich and complex problem. For the case of Lagrangian tracers, the multi-scale nature of turbulence is responsible for the rapid super diffusion of fluid parcels. We have shown that super-diffusion can be trivially understood as a simple cascade of successive scale dependent ballistic steps, what offers a much simpler paradigm than the original conjecture by Richardson of a scale dependent diffusive process. Turbulent super-diffusion directly impacts for instance the way a cloud of particles disperses from a point source. The rapid growth of particles separation ensures a rapid growth of the cloud towards large scales, where efficient turbulent mixing by uncorrelated eddies eventually disperses the particles much more effectively than molecular diffusion. But at the same time this rapid growth results in important heterogeneities of the distribution of particles at intermediate (inertial) scales, with regions of local high concentration, separated by depleted regions (see for instance Fig. 2a.4). Practical consequences are potentially important, if one considers for instance the release of hazardous substances (as pollutants or contaminants): although far from the source the substance will disperse by the uncorrelated turbulent mixing, in the vicinity of the pollution source, puffs of high concentration can exist intermittently, where although the overall average concentration may remain below a given toxicity threshold, local and instantaneous levels can exceed the admissible limit. Such source of heterogeneities in the concentration field of dispersed particles is entirely related to the multi scale nature of turbulence and does not require particles to have inertia. It is only visible though if particles are dispersed from a source initially at intermediate (inertial) scales. If particles are released from an initial uniformly distribution at large scales, no such heterogeneities will appear, as a consequence of the conservative nature of particles dynamics (using the terminology of dynamical systems). This contrasts with the case of particles with inertia, which will experience preferential concentration and hence the emergence of heterogeneities with persistent dense regions and depleted regions even if they are initially released from a uniform distribution. The emergence of such heterogeneities is believed to play an important role in many natural and industrial systems. For instance, preferential concentration of inertial particles may enhance the probability of collision (and hence of coalescence) of water droplets in clouds, and hence play an important role in rain initiation. Similarly it is believed that it can promote the agglomeration of fine particles in accretion disks, hence accelerating the formation of planetesimals. In industrial applications, it can for instance play a role in the coalescence of fuel droplets in diesel engines for instance, and therefore affect the energetic efficiency, with important economical and environmental implications.

References

A Aliseda, A Cartellier, F Hainaux, and J C Lasheras. Effect of preferential concentration on the settling velocity of heavy particles in homogeneous isotropic turbulence. Journal of Fluid Mechanics, 468:77–105, 2002.

Alfred Barnard Basset. A treatise on hydrodynamics, volume 2. Deighton, Bell and Co., 1888.

G K Batchelor. The application of the similarity theory of turbulence to atmospheric diffusion. *Quarterly Journal of the Royal Meteorological Sociey*, 76(328):133–146, 1950. ISSN 0035-9009.

G K Batchelor. Diffusion in a field of homogeneous turbulence II. The relative motion of particles. Proceedings of the Cambridge Philosophical Society, 48(2):345–362, 1952.

Jeremie Bec, Luca Biferale, Massimo Cencini, Alessandra Lanotte, Stefano Musacchio, and Federico Toschi. Heavy particle concentration in turbulence at dissipative and inertial scales. *Physical Review Letters*, 98(8):84502, 2007. ISSN 0031-9007. doi:10.1103/PhysRevLett.98.084502. URL https://doi.org/10.1103/PhysRevLett.98.084502.

Jacob Berg, Beat Lüthi, Jakob Mann, and Sø ren Ott. Backwards and forwards relative dispersion in turbulent flow: An experimental investigation. *Physical Review E*, 74(1):016304, July 2006. ISSN 1539-3755. doi:10.1103/PhysRevE.74.016304. URL http://link.aps.org/doi/10.1103/PhysRevE.74.016304.

L Biferale, G Boffetta, A Celani, B J Devenish, A Lanotte, and F Toschi. Lagrangian statistics of particles pairs in homogeneous isotropic turbulence. Physics of Fluids, 17:115101, 2005.

Rehab Bitane, Holger Homann, and Jérémie Bec. Time scales of turbulent relative dispersion. *Physical Review E*, 86(4):045302, October 2012. ISSN 1539-3755. doi:10.1103/PhysRevE.86.045302. URL http://link.aps.org/doi/10.1103/PhysRevE.86.045302.

G Boffetta and A Celani. Pair dispersion in turbulence. *Physica A*, 280(1-2):1–9, 2000. ISSN 0378-4371.

G. Boffetta and I. Sokolov. Relative Dispersion in Fully Developed Turbulence: The Richardsons Law and Intermittency Corrections. *Physical Review Letters*, 88(9):094501, February 2002. ISSN 0031-9007. doi:10.1103/PhysRevLett.88.094501. URL http://link.aps.org/doi/10.1103/PhysRevLett.88.094501.

Mickael Bourgoin. Turbulent pair dispersion as a ballistic cascade phenomenology. Journal of Fluid Mechanics, 772:678–704, 2015.

Mickaël Bourgoin and Haitao Xu. Focus on dynamics of particles in turbulence. *New Journal of Physics*, 16(8):085010, August 2014. ISSN 1367-2630. doi:10.1088/1367-2630/16/8/085010. URL http://stacks.iop.org/1367-2630/16/i=8/a=085010?key=crossref.963834ca83607a6aa71791a4a68b0777.

Mickael Bourgoin, Nicholas T Ouellette, Haitao Xu, Jacob Berg, and Eberhard Bodenschatz. The role of pair dispersion in turbulent flow. *Science*, 311(5762):835–838, 2006. doi:10.1126/science.1121726.

Mickael Bourgoin, Jean-François Pinton, and Romain Volk. Lagrangian methods in experimental fluid mechanics. In Thomas von Larcher and Paul D. Williams, editors, *Modeling Atmospheric and Oceanic Flows: Insights from Laboratory Experiments and Numerical Simulations*, page 360. John Wiley & Sons, Inc., 2014. ISBN 1118855930. URL https://books.google.com/books?hl=fr&lr=&id=8GezBQAAQBAJ&pgis=1.

Joseph Boussinesq. *Th{é}orie analytique de la chaleur: mise en harmonie avec la thermodynamique et avec la th{é}orie m{é}canique de la lumi{é}re*, volume 2. Gauthier-Villars, 1903.

S W Coleman and J C Vassilicos. A unified sweep-stick mechanism to explain particle clustering in two- and three-dimensional homogeneous, isotropic turbulence. Physics of Fluids, 21(11):113301, November 2009. ISSN 1070-6631. doi:10.1063/1.3257638.

T. Faber and J. C. Vassilicos. Turbulent pair separation due to multiscale stagnation point structure and its time asymmetry in two-dimensional turbulence. *Physics of Fluids*, 21(1):015106, 2009. ISSN 10706631. doi:10.1063/1.3059618. URL http://scitation.aip.org/content/aip/journal/pof2/21/1/10.1063/1.3059618.

Jarai-Szabo Ferenc and Zoltan Neda. On the size distribution of Poisson Voronoi cells. Physica A-Statistical Mechanics and Its Applications, 385(2):518–526, 2007. doi:10.1016/j.physa.2007.07.063.

Jr Fessler, Jd Kulick, and Jk Eaton. Preferential Concentration Of Heavy-Particles In A Turbulent Channel Flow. *Physics of Fluids*, *6*(11):3742–3749, 1994.

Renée Gatignol. The Faxen formulas for a rigid particle in an unsteady non-uniform Stoke flow. Journal de Mécanique Théorique et Appliquée, 2(2):143–160, 1983.

Susumu Goto and J. Vassilicos. Sweep-Stick Mechanism of Heavy Particle Clustering in Fluid Turbulence. *Physical Review Letters*, 100(5):54503, February 2008. ISSN 0031-9007. doi:10.1103/PhysRevLett.100.054503. URL http://link.aps.org/doi/10.1103/PhysRevLett.100.054503.

M Huber, J C McWilliams, and M Ghil. A climatology of turbulent dispersion in the troposphere. Journal of the Atmospheric Sciences, 58(16):2377–2394, 2001.

Jennifer Jucha, Haitao Xu, Alain Pumir, and Eberhard Bodenschatz. Time-reversal-symmetry Breaking in Turbulence. *Physical Review Letters*, 113(5):054501, July 2014. ISSN 0031-9007. doi:10.1103/PhysRevLett.113.054501. URL http://link.aps.org/doi/10.1103/PhysRevLett.113.054501.

J. Klafter, A. Blumen, and M. F. Shlesinger. Stochastic pathway to anomalous diffusion. Physical Review E, 35:3081–3085, 1987.

A. Kolmogorov. The local structure of turbulence in incompressible viscous fluid for very large Reynolds numbers. *Dokl. Akacl. Nauk SSSR*, 30:301–305, 1941. ISSN 0002-3264.

Arthur LaPorta, Greg A Voth, Alice M Crawford, Jim Alexander, and Eberhard Bodenschatz. Fluid particle accelerations in fuly developped turbulence. Nature, 409:1017, 2001.

J Mann, S Ott, and J S Andersen. Experimental study of relative, turbulent diffusion. Technical Report Riso-R-1036 (EN), Risoe National Laboratory, Roskilde, Denmark, 1999.

M Maxey. The Gravitational Settling Of Aerosol-Particles In Homogeneous Turbulence And Random Flow-Fields. Journal of Fluid Mechanics, 174:441–465, 1987.

Martin R Maxey and James J Riley. Equation of motion for a small rigid sphere in a nonuniform flow. Physics of Fluids, 26(4):883–889, 1983.

B. Mehlig, M. Wilkinson, K. Duncan, T. Weber, and M. Ljunggren. Aggregation of inertial particles in random flows. *Physical Review E*, 72(5):051104, November 2005. ISSN 1539-3755. doi:10.1103/PhysRevE.72.051104. URL http://link.aps.org/doi/10.1103/PhysRevE.72.051104.

Bernhard Mehlig and Michael Wilkinson. Coagulation by Random Velocity Fields as a Kramers Problem. *Physical Review Letters*, 92(25):250602, June 2004. ISSN 0031-9007. doi:10.1103/PhysRevLett.92.250602. URL http://link.aps.org/doi/10.1103/PhysRevLett.92.250602.

R Monchaux, M Bourgoin, and A Cartellier. Preferential concentration of heavy particles: A Voronoi analysis. Physics of Fluids, 22(10):103304, 2010. doi:10.1063/1.3489987.

Romain Monchaux, Mickael Bourgoin, and Alain Cartellier. Analyzing preferential concentration and clustering of inertial particles in turbulence. *International Journal of Multiphase Flow*, 40:1–18, 2012. ISSN 0301-9322. doi:10.1016/j.ijmultiphaseflow.2011.12.001. URL http://www.sciencedirect.com/science/article/pii/S030193221100245X.

M NELKIN and R M KERR. DECAY OF SCALAR VARIANCE IN TERMS OF A MODIFIED RICHARDSON LAW FOR PAIR DISPERSION. *Physics of Fluids*, 24(9):1754–1756, 1981.

Martin Obligado, Tomas Teitelbaum, Alain Cartellier, Pablo D. Mininni, and Mickael Bourgoin. Preferential Concentration of Heavy Particles in Turbulence. Journal of Turbulence, 15(5):293–310, 2014.

Piero Olla. Preferential concentration versus clustering in inertial particle transport by random velocity fields. *Physical Review E*, 81(1):16305, January 2010. ISSN 1539-3755. doi:10.1103/PhysRevE.81.016305. URL http://link.aps.org/doi/10.1103/PhysRevE.81.016305.

Carl Wilhelm Oseen. *{Ü}ber die Stoke'sche Formel und {ü}ber eine verwandte Aufgabe in der Hydrodynamik:...* Almqvist & Wiksell, 1911.

Soren Ott and Jakob Mann. An experimental investigation of the relative diffusion of particle pairs in three-dimensional turbulent flow. Journal of Fluid Mechanics, 422:207–223, 2000.

Nicholas T Ouellette, Haitao Xu, Mickaël Bourgoin, and Eberhard Bodenschatz. An experimental study of turbulent relative dispersion models. *New Journal of Physics*, 8(6):109, June 2006. ISSN 1367-2630. doi:10.1088/1367-2630/8/6/109. URL http://stacks.iop.org/1367-2630/8/i=6/a=109?key=crossref.6c8e470b5e7116e5021ffd8b32247aa7.

Lewis F Richardson. Atmospheric Diffusion Shown on a Distance-Neighbour Graph. *Proceedings of the Royal Society of London, Series A*, 110(756):709–737, 1926.

Juan P L C Salazar and Lance R Collins. Two-Particle Dispersion in Isotropic Turbulent Flows. *ANNUAL REVIEW OF FLUID MECHANICS*, 41:405–432, 2009. ISSN 0066-4189. doi:10.1146/annurev.fluid.40.111406.102224.

Juan P L C Salazar, Jeremy de Jong, Lujie Cao, Scott H Woodward, Hui Meng, and Lance R Collins. Experimental and numerical investigation of inertial particle clustering in isotropic turbulence. *Journal of Fluid Mechanics*, 600(1):245–256, 2008. ISSN 0022-1120. doi:10.1017/S0022112008000372. URL http://journals.cambridge.org/production/action/cjoGetFulltext?fulltextid=1818596.

Ewe Wei Saw, Raymond A Shaw, Sathyanarayana Ayyalasomayajula, Patrick Y Chuang, and Armann Gylfason. Inertial clustering of particles in high-reynolds-number turbulence. *Physical Review Letters*, 100(21), 2008. ISSN 0031-9007. doi:10.1103/PhysRevLett.100.214501.

Brian Sawford. Turbulent Relative Dispersion. Annual Review of Fluid Mechanics, 33:289–317, 2001.

Brian L. Sawford, P. K. Yeung, and Michael S. Borgas. Comparison of backwards and forwards relative dispersion in turbulence. *Physics of Fluids*, 17(9):095109, 2005. ISSN 10706631. doi:10.1063/1.2047593. URL http://scitation.aip.org/content/aip/journal/pof2/17/9/10.1063/1.2047593.

Stephen J Scott, Aditya U Karnik, and John S Shrimpton. On the quantification of preferential accumulation. *INTERNATIONAL JOURNAL OF HEAT AND FLUID FLOW*, 30(4):789–795, 2009. ISSN 0142-727X. doi:10.1016/j.ijheatfluidflow.2009.02.002.

K D Squires and J K Eaton. Preferential concentration of particles by turbulence. Physics of Fluids A, 3(5):1169–1178, 1991.

Katepalli R Sreenivasan. On the universality of the Kolmogorov constant. Physics of Fluids, 7(3):2778–2784, 1995.

G I Taylor. Diffusion by continuous movements. Proc. Lond. Math. Soc., 20:196–212, 1922.

Federico Toschi and Eberhard Bodenschatz. Lagrangian Properties of Particles in Turbulence. Annual Review of Fluid Mechanics, 41:375–404, 2009. doi:10.1146/annurev.fluid.010908.165210.

Greg A Voth, Arthur LaPorta, Alice M Crawford, Jim Alexander, and Eberhard Bodenschatz. Measurement of particle accelerations in fully developed turbulence. Journal of Fluid Mechanics, 469:121–160, 2002.

A M Wood, W Hwang, and J K Eaton. Preferential concentration of particles in homogeneous and isotropic turbulence. International Journal of Multiphase Flow, 31:1220–1230, 2005.

P K Yeung and S B Pope. Lagrangian statistics from direct numerical simulations of isotropic turbulence. Journal of Fluid Mechanics, 207:531–586, 1989.

Collective Dynamics of Particles in Viscous Flows with an Emphasis on Slender Rods

Jason E. Butler

Abstract Basic principles that govern the viscous motion of non-colloidal particles are described, and then the principles are applied to the analysis and simulation of the collective motion of particles in a concentrated suspension. Though rigid spheres are discussed in general, the dynamics of rigid rods are the focus of the given examples, equations, and simulation methods.

1 Introduction

The accurate and reliable prediction of the collective motion of particles suspended in fluids remains one of the key challenges of fluid mechanics, despite the widespread importance of the topic to many industrial processes, such as mixing and formation of composites, and to understanding natural phenomena, such as the creation of deltas from the sedimentation of river silt (Larson 1999). Even the qualitative properties of suspensions under flow continue to defy predictions in many cases (Butler 2014).

For predicting dynamics, accurate continuum-level equations would be most useful due to the scales and complexities of realistic flows that are of most interest, but these equations are not well-established for even relatively simple cases. The continuing development of accurate conceptual and numerical models requires careful comparisons between experimental observations and simulations based upon rigorously applied theories. The development of this latter approach for the purposes of resolving questions concerning the collective motion of particles, with a focus on rods, is the subject of this chapter.

By necessity, the conditions of the suspensions must be restricted to make some progress; here, most of the assumptions made will be the simplest possible. The suspending fluid is incompressible, Newtonian, and viscous. For sufficiently large

J.E. Butler (✉)
Department of Chemical Engineering, University of Florida,
Gainesville, FL 32611, USA
e-mail: butler@che.ufl.edu
URL: http://www.che.ufl.edu/faculty/butler

C. Marchioli (ed.), *Collective Dynamics of Particles*,
CISM International Centre for Mechanical Sciences 576,
DOI 10.1007/978-3-319-51226-6_4

viscosity or small particle size and flow rates, the nonlinear and time-dependent contributions in the Navier–Stokes equations become insignificant, and the fluid flow is governed by the Stokes equations. Hence, the fluid motion is non-inertial, and the same assumption is made for the particle motion. The particle and fluid motion match at every point on the surface of the particles, which are considered rigid for all the work discussed here. The particles are also relatively large so that colloidal effects, such as Brownian motion, are insignificant.

The general principles that govern the motion of the particles are described in Sect. 2, and concepts that influence the collective motion of the particles, along with some examples from the motion of spheres, are introduced (Sect. 3). The equations of motion for a rigid, slender body are given in Sect. 4 along with example calculations that demonstrate key features of the flows. Using the ideas of Sect. 4, a simulation method for the dynamics of a collection of rods is developed in Sect. 5. Some examples of calculations using the method and some concluding remarks are given in Sects. 6 and 7.

2 General Principles

A number of consequences for the allowable motions of the particles follows from the stated assumptions regarding the properties of the fluid, particles, and rates of flow.[1] These consequences enable a number of simplifications that can be used to advantage when constructing simulation methods and can also be used to evaluate the plausibility of the resulting solutions.

Because of the linearity of Stokes equations, the velocities in a fluid flow can be superimposed. Figure 1 shows the example of a simple shear flow, which can be decomposed into a rotational flow and an extensional flow. Likewise, the flow fields reverse upon reversing the forcings, as also illustrated in Fig. 1 for flow through a constriction, as a consequence of the linearity of Stokes equations. Both of these consequences are useful simplifications for the purposes of calculating the fluid flows due to the presence of multiple particles, where the velocities caused by forces acting on the particles can be calculated independently and then added.

Additional, general consequences of the assumptions for the motion of the particles can be determined from the assumptions. Ignoring inertia, and hence acceleration, of the particle means that the particle is always moving at its terminal velocity. As the forces or velocities acting on the particle change in time (or position, as the particle location changes in time), the particle accelerates to the new velocity infinitely fast. This, together with the instantaneity of the fluid flow at all length scales at low Reynolds number, indicates that the particle velocities can be calculated from the instantaneous positions and forces, without needing to know the history of the velocities of either the particles or the fluid. This results in the significant simplifi-

[1] A thorough and readable introduction to these concepts is available in the text by Guazzelli and Morris (2012). Also, the text by Kim and Karrila (2005) is a useful handbook containing multiple relationships for performing calculations on these types of suspensions.

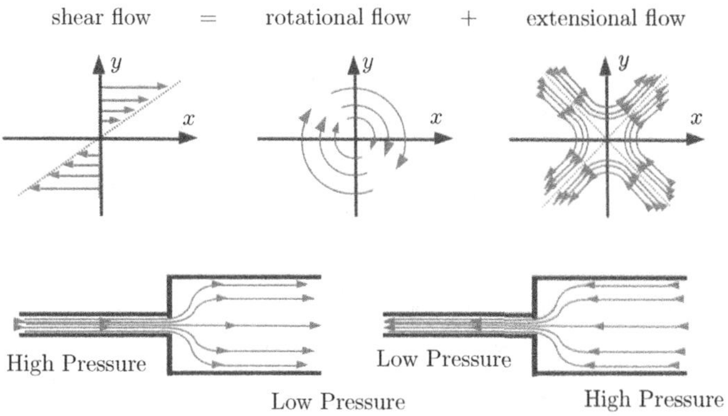

Fig. 1 Linearity of the flows enables superposition of the velocity fields and results in flow solutions that are reversible. For example (*top*), the shearing flow of a viscous fluid can be written as the sum, in equal parts, of a rotational and planar extension flow. On the *bottom*, the reversibility of the flow through a constriction upon reversing the pressure differential is illustrated

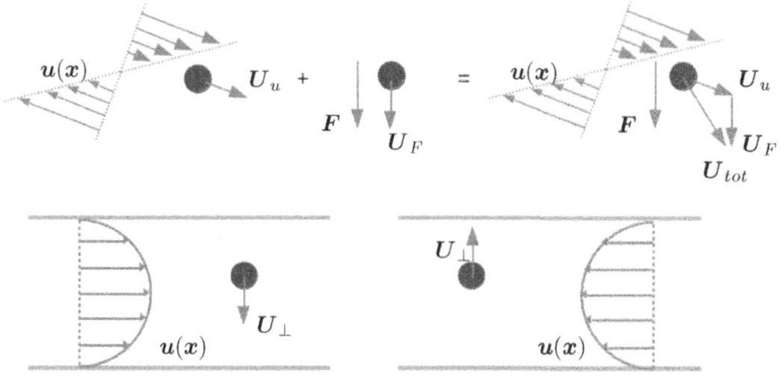

Fig. 2 Owing to the lack of inertia, the net motion $\boldsymbol{U}_{tot}$ (*top*) of a particle at any moment in time due to a flow field $\boldsymbol{u}(\boldsymbol{x})$ and a force $\boldsymbol{F}$ is the sum of the motions caused by each when acting alone. Reversibility (*bottom*) requires that the transverse motion $U_\perp$ of a spherical particle in a parabolic flow field equal zero since reversing the flow direction would require that the velocity reverse sign

cation illustrated in Fig. 2 that the net motion of a particle due to multiple forces and flows can be calculated at any moment in time by adding the steady motions caused by each force and flow.

Analogous to the reversible properties of the fluid flow, the motion of a particle subject to all of the restrictions listed above must also be reversible. This property can be useful in identifying the allowable motions of particles. Figure 2 illustrates the example of a spherical particle suspended in a parabolic flow. Upon reversing the direction of the flow field, all velocity components for the particle must also reverse, including any velocity transverse ($U_\perp$) to the flow field. Hence, $U_\perp$ must be zero since the particle is symmetric (i.e. has no mechanism for breaking symmetry upon flow reversal).

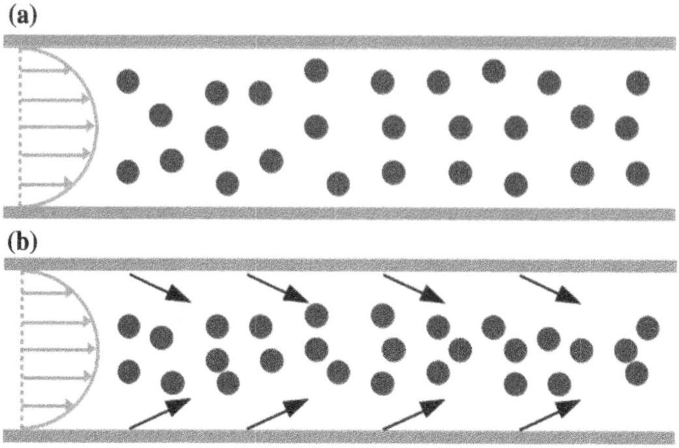

Fig. 3 An initially uniform suspension of spheres **a** transported through a tube due to a pressure-driven flow migrates toward the centerline, leaving **b** a depletion region near the bounding walls. Reversing the direction of the flow does not remix the suspension

The assumptions and associated consequences can be used to evaluate the allowable motions of particles, but the collective motion of concentrated suspensions seem to violate many of the principles. For example, experiments demonstrate that shearing a suspension of spheres can drive the particles from regions of high shear rate toward regions of low shear rate. This phenomena was identified clearly by Leighton and Acrivos (1987), though evidence of such concentration inhomogeneities in flowing suspensions dates back to Karnis et al. (1966). Consider the specific case of the transport of a concentrated suspension of spheres through a tube due to a pressure gradient as shown in Fig. 3. Several studies (Altobelli et al. 1991; Hampton et al. 1997; Butler and Bonnecaze 1991) have verified that an initially uniform suspension of spheres will migrate to the center of the pipe, even at very low values of the Reynolds number; simulations demonstrate the same (Nott and Brady 1994; Yeo and Maxey 2011).

3 Origins of Collective Dynamics

The irreversible behavior illustrated in Fig. 3 is at odds with the arguments made in Fig. 2: even though the fluid motion is linear and the motion of a single particle is reversible, the evidence demonstrates that the collective dynamics of the particles can be complex. There are at least two possible explanations. One is that the system of concentrated particles is chaotic. As a result, the collective motion of the particles is irreversible in practice, despite the fact that the governing equations are formally reversible. The other is that additional forces, such as particle-particle contacts are occurring in the system.

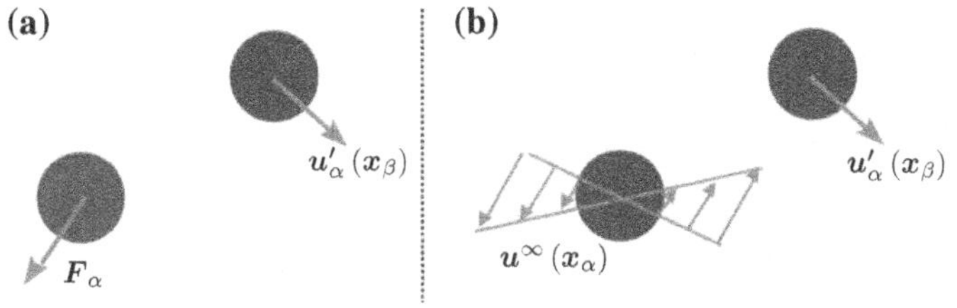

Fig. 4 Forces (**a**) or straining flows (**b**) on a particle α create an additional velocity (u'_α) on other particles, such as particle β which is located at x_β

In fact, evidence indicates that both of these phenomena contribute to the loss of reversibility in the collective dynamics of suspensions, where the dominant source depends upon the specific situation. In the following, the hydrodynamic origins of chaos are explored using a point-particle model (Sect. 3.1) of sedimenting spheres and some additional discussion of the roles of particle contacts is given in Sect. 3.2.

3.1 Chaotic Motion Due to Hydrodynamic Interactions

One argument is that the collective motion of the suspensions, despite being governed by reversible equations, are sensitive to small disturbances that magnify in time; in other words, the particle motions are chaotic. The non-linearity, which must be present in a chaotic system, is due to the hydrodynamic interactions between the particles. As illustrated in Fig. 4, applying a force or a straining flow to a particle results in a disturbance to the fluid velocity. These disturbance velocities alter the motion of the other particles in suspension, depending on the relative positions. These hydrodynamic interactions are non-linear and can be chaotic.

Here, a model for spheres falling through a viscous fluid is constructed as a demonstration that the collective motion is chaotic due to the long-range hydrodynamic interactions. The spheres are assumed to be small in comparison to their separation distance and are far from any boundaries. In this point-particle limit, the rotational motion is ignored, and only the leading order contributions are retained. The center-of-mass motion of a spherical particle is governed by a force balance,

$$m\ddot{x} = F_D + \Sigma F, \tag{1}$$

where m is the mass and $\ddot{x}$ is the acceleration of the particle which is located at a position x. The left-hand side of this equation is set to zero, as the particle is non-inertial. On the right-hand side, the drag force on the particle, F_D, has been separated from the sum of other forces, ΣF, that may act on the particle.

The drag force is proportional, in this limit of low Reynolds number, to the difference in velocity of the particle, $\dot{x}$, and the fluid velocity at the position of the particle, $u(x)$,

$$F_D = -\mathcal{R} \cdot (\dot{x} - u(x)). \tag{2}$$

The proportionality is given by the resistance matrix, $\mathcal{R}$. Solving for the velocity gives

$$\dot{\boldsymbol{x}} = \mathcal{M} \cdot (\Sigma \boldsymbol{F}) + \boldsymbol{u}(\boldsymbol{x}), \tag{3}$$

where $\mathcal{M} = \mathcal{R}^{-1}$ is the mobility of the particle. For spherical particles, the mobility is inversely proportional to the viscosity μ of the fluid, the particle radius a, and is diagonal,

$$\mathcal{M} = \frac{1}{6\pi\mu a}\boldsymbol{I}. \tag{4}$$

As a result, the particle motion (in the absence of fluid flows) is in the same direction as the resultant forces,

$$\boldsymbol{U} = \frac{1}{6\pi\mu a}(\Sigma \boldsymbol{F}), \tag{5}$$

and is instantaneous: as the forces change in time, so does the velocity.

To calculate the collective motion of a group of particles, the fluid velocity appearing in Eq. 3 must include the fluid disturbance of each particle on the others. For any particular particle α located at $\boldsymbol{x}_\alpha$, the velocity is given by

$$\dot{\boldsymbol{x}}_\alpha = \mathcal{M}_\alpha \cdot (\Sigma \boldsymbol{F})_\alpha + \boldsymbol{u}^\infty(\boldsymbol{x}_\alpha) + \sum_\beta \boldsymbol{u}'_\beta(\boldsymbol{x}_\alpha), \tag{6}$$

where the forces $(\Sigma \boldsymbol{F})_\alpha$ and mobility $\mathcal{M}_\alpha$ can depend on the particle α. The velocity of the fluid, $\boldsymbol{u}(\boldsymbol{x}_\alpha)$, is the sum of the fluid velocity in the absence of any particles, $\boldsymbol{u}^\infty(\boldsymbol{x}_\alpha)$, and the disturbances, $\boldsymbol{u}'_\beta(\boldsymbol{x}_\alpha)$, from all other particles β in the fluid. For a particle β acted upon by forces $\Sigma \boldsymbol{F}_\beta$, the leading contribution to the velocity at $\boldsymbol{x}_\alpha$ is

$$\begin{aligned} \boldsymbol{u}'_\beta(\boldsymbol{x}_\alpha) &= \frac{1}{8\pi\mu\left|\boldsymbol{x}_\alpha - \boldsymbol{x}_\beta\right|}\left(\boldsymbol{I} + \frac{(\boldsymbol{x}_\alpha - \boldsymbol{x}_\beta)(\boldsymbol{x}_\alpha - \boldsymbol{x}_\beta)}{\left|\boldsymbol{x}_\alpha - \boldsymbol{x}_\beta\right|^2}\right) \cdot (\Sigma \boldsymbol{F}_\beta) \\ &= \boldsymbol{G}(\boldsymbol{x}_\alpha - \boldsymbol{x}_\beta) \cdot (\Sigma \boldsymbol{F}_\beta), \end{aligned} \tag{7}$$

where $\left|\boldsymbol{x}_\alpha - \boldsymbol{x}_\beta\right|$ is the magnitude of $\boldsymbol{x}_\alpha - \boldsymbol{x}_\beta$. The Oseen tensor, $\boldsymbol{G}(\boldsymbol{x}_\alpha - \boldsymbol{x}_\beta)$, is the Green's function for Stokes equations in an unbounded fluid where the velocity disturbance decays to zero at infinity (Guazzelli and Morris 2012; Kim and Karrila 2005). Figure 5 illustrates the flow field given by Eq. 7.

As a result of the disturbance velocities, the solution for the position of a particle α in time,

$$\boldsymbol{x}_\alpha(t) = \int_{t'=0}^{t} \dot{\boldsymbol{x}}_\alpha\left(\boldsymbol{x}_\alpha(t'), \boldsymbol{x}_\beta(t')\right) dt' + \boldsymbol{x}_\alpha(0), \tag{8}$$

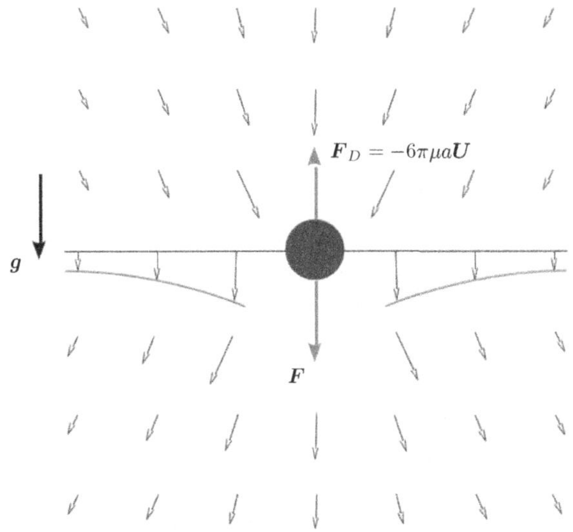

Fig. 5 Illustration of the flow disturbance caused by a force $\boldsymbol{F}$, due to gravity $\boldsymbol{g}$ for example. The force is balanced by the drag force $\boldsymbol{F}_D$ which is proportional to the velocity of the particle, $\boldsymbol{U}$. The motion produces a flow field (*blue arrows*) of the form shown. (color figure online)

depends upon the relative positions of all the particles in the suspension. This multi-body problem for the positions is non-linear, owing to the non-linear dependence of the interactions as defined in Eq. 7, and can be chaotic.

For two particles suspended in a quiescent fluid of infinite extent, only the motion due to the forces and the disturbance of each particle on the other needs to be considered as shown in Fig. 6. The motion is then fully described by

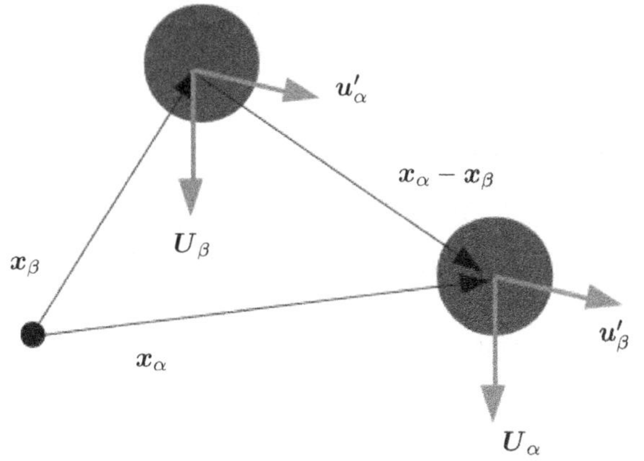

Fig. 6 Two particles sedimenting due to a force in an infinite fluid in the absence of any externally imposed flow ($\boldsymbol{u}^\infty = 0$). If the radius of each particle and the force on each particle is the same, then $\boldsymbol{U}_\alpha = \boldsymbol{U}_\beta$ and $\boldsymbol{u}'_\alpha = \boldsymbol{u}'_\beta$

$$\dot{\boldsymbol{x}}_\alpha = \boldsymbol{U}_\alpha + \boldsymbol{u}'_\beta(\boldsymbol{x}_\alpha) \tag{9}$$

$$\dot{\boldsymbol{x}}_\beta = \boldsymbol{U}_\beta + \boldsymbol{u}'_\alpha(\boldsymbol{x}_\beta), \tag{10}$$

or more clearly,

$$\dot{\boldsymbol{x}}_\alpha = \frac{1}{6\pi\mu a}\boldsymbol{F}_\alpha + \boldsymbol{G}(\boldsymbol{x}_\alpha - \boldsymbol{x}_\beta)\cdot\boldsymbol{F}_\beta \tag{11}$$

$$\dot{\boldsymbol{x}}_\beta = \frac{1}{6\pi\mu a}\boldsymbol{F}_\beta + \boldsymbol{G}(\boldsymbol{x}_\beta - \boldsymbol{x}_\alpha)\cdot\boldsymbol{F}_\alpha. \tag{12}$$

The Oseen matrix is a symmetric function of the separation of the particles (i.e. $\boldsymbol{G}(\boldsymbol{x}_\alpha - \boldsymbol{x}_\beta) = \boldsymbol{G}(\boldsymbol{x}_\beta - \boldsymbol{x}_\alpha)$), so the motion of the particles is identical if the two particles have the same properties (radius) and are acted upon by the same force. The relative positions remain fixed in time,

$$\boldsymbol{x}_\beta(t) - \boldsymbol{x}_\alpha(t) = \boldsymbol{x}_\beta(0) - \boldsymbol{x}_\alpha(0), \tag{13}$$

and hence the collective motion of two particles is not chaotic. The interactions do alter the motion, though. A pair of spheres falls faster than a single sphere, regardless of the relative positions for a finite separation distance; the velocity is largest for spheres aligned with the direction of force ($[\boldsymbol{x}_\alpha - \boldsymbol{x}_\beta]\cdot\boldsymbol{F}$ is maximized) and smallest for spheres aligned perpendicular to the force ($[\boldsymbol{x}_\alpha - \boldsymbol{x}_\beta]\cdot\boldsymbol{F} = 0$). Also, the pair of spheres can drift in directions perpendicular to the force, whereas a single sphere can not in this limit of zero Reynolds number.

The alteration of the dynamics of particles due to collective motion is most often expressed as a change in the mobility. To make that more clear, the equations for the pair of particles is written in a more convenient format,

$$\begin{bmatrix}\dot{\boldsymbol{x}}_\alpha \\ \dot{\boldsymbol{x}}_\beta\end{bmatrix} = \boldsymbol{M}\cdot\begin{bmatrix}\boldsymbol{F}_\alpha \\ \boldsymbol{F}_\beta\end{bmatrix}, \tag{14}$$

where the mobility is given by

$$\boldsymbol{M} = \begin{bmatrix}\frac{1}{6\pi\mu a}\boldsymbol{I} & \boldsymbol{G}(\boldsymbol{x}_\alpha - \boldsymbol{x}_\beta) \\ \boldsymbol{G}(\boldsymbol{x}_\beta - \boldsymbol{x}_\alpha) & \frac{1}{6\pi\mu a}\boldsymbol{I}\end{bmatrix}. \tag{15}$$

The solution for the dynamics of a larger number of N particles undergoing motion due to body forces (such as gravity) can be generated by generalizing the mobility

$$\boldsymbol{M} = \begin{bmatrix}\mathcal{M}_{11} & \mathcal{G}_{12} & \cdots & \mathcal{G}_{1N} \\ \mathcal{G}_{21} & \mathcal{M}_{22} & \cdots & \cdots \\ \cdots & \cdots & \cdots & \cdots \\ \mathcal{G}_{N1} & \cdots & \cdots & \mathcal{M}_{NN}\end{bmatrix}, \tag{16}$$

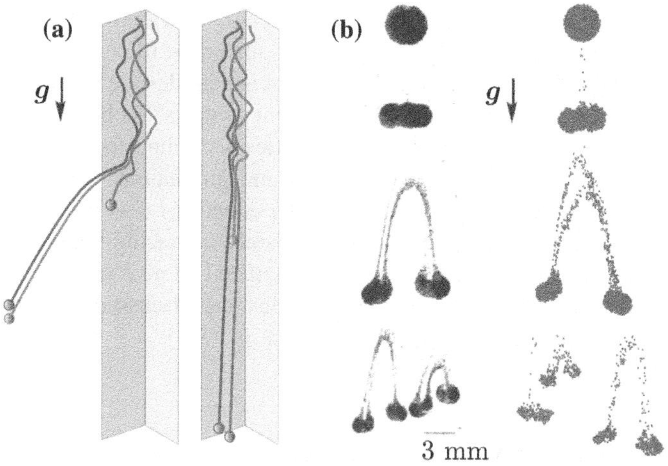

Fig. 7 Examples of the chaotic motion of sedimenting spheres. **a** Small perturbations in the initial placement of three spheres produces very different results as shown by the comparison of two such trajectories. **b** Experiments (*left* in *black*) and simulations (*right* in *red*) on a sedimenting cloud of spheres exhibit chaotic dynamics (Images are courtesy of Bloen Metzger, Aix-Marseille Université, France). (color figure online)

where

$$\mathcal{M}_{ii} = \frac{1}{6\pi\mu a}\boldsymbol{I} \tag{17}$$

and

$$\mathcal{G}_{ij} = \mathcal{G}\left(\boldsymbol{x}_i - \boldsymbol{x}_j\right). \tag{18}$$

Though the above equations represent a simplification of the relevant physics, they are sufficient to demonstrate that hydrodynamic interactions drive a chaotic motion. Examining the motion of three particles using an algorithm similar to that given above, Janosi et al. (1997) showed that small perturbations in the initial positions of the particles resulted in exponentially diverging trajectories; an example is shown in Fig. 7a. Such motions are a direct evidence of chaos. For a large number of particles that are initially placed in a spherical cloud as shown in Fig. 7b, simulations using methods identical to those explained here predict a complex set of dynamics (Metzger et al. 2007). Experiments of the patterns of the evolving cloud of sedimenting particles are also shown in Fig. 7b as a comparison.

3.2 Irreversibilities Due to Contact Interactions

For the shearing flows of concentrated suspensions of spherical particles, studies (Metzger and Butler 2010; Metzger et al. 2013) indicate that the hydrodynamic interactions, which are weaker in this case, are not sufficiently chaotic to explain observations of irreversibility made in experiments on oscillating systems of particles (Pine et al. 2005). Hence, another argument is that, despite the restrictions, the lubricating layer between particles fails to prevent contacts. Such particle-to-particle contacts break the reversibility of the trajectories according to available evidence (Pham et al. 2016) that compares experiments and simulations. Those simulations, by necessity, include terms that are not in the equations developed in the previous section. Chiefly, corrections for hydrodynamic interactions at short-range, or so-called lubrication interactions, must be calculated to give accurate trajectories of the particles. This issue, and other corrections, are discussed thoroughly in the literature of Stokesian dynamics (Brady and Bossis 1988).

4 Calculating the Motion of Rods

To address questions regarding whether contact interactions and chaos also influence the collective motion of more general suspensions, equations are developed for simulating the motion of rigid rods. As in Eqs. 1–3, the motion of a rigid particle suspended in a fluid is given by solving force and torque balances, and the inertia of the particle is assumed to be negligible. Consequently, the balances reduce to

$$\boldsymbol{F}_D = -\Sigma \boldsymbol{F} \tag{19}$$

and

$$\boldsymbol{\tau}_D = -\Sigma \boldsymbol{\tau}, \tag{20}$$

where $\boldsymbol{F}_D$ and $\boldsymbol{\tau}_D$ are the hydrodynamic forces and torques on the particle and $\Sigma \boldsymbol{F}$, and $\Sigma \boldsymbol{\tau}$ are the sums of the other forces and torques acting on the particles.

The drag force $\boldsymbol{F}_D$ is proportional to the difference of the particle and fluid velocity, and so consequently Eq. 19 can be solved to give the center-of-mass motion of a particle. Likewise, the angular motion is given by solving Eq. 20 since $\boldsymbol{\tau}_D$ is proportional to the difference of the particle and fluid angular velocity. The next section reviews possible models for the drag on rod-like particles.

4.1 Hydrodynamic Models for Rigid Fibers

A number of models have been used in the literature for representing the viscous motion of a rigid rod; four examples are shown in Fig. 8. Each model incorporates a different level of approximation for the motion, with increasing accuracy generally resulting in the complexity of the equations.

Perhaps the simplest of models consists of two spheres, or beads, connected by a rigid constraint. The constraining rod pictured in Fig. 8 generates no hydrodynamic

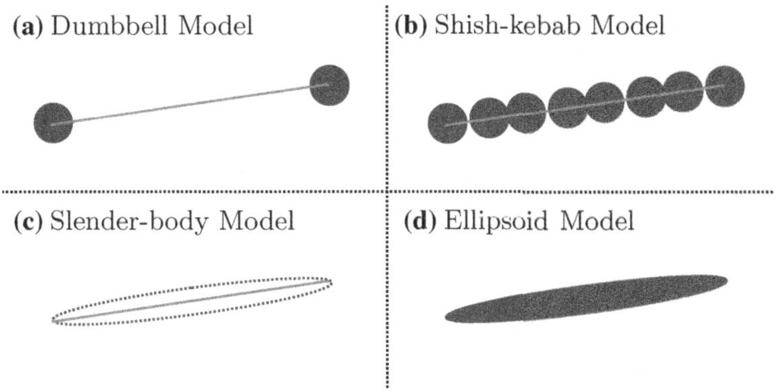

Fig. 8 Four physical models used for calculating the motion of rigid rods. **a** The dumbbell model consists of two beads connected rigidly at a fixed distance. **b** The "shish-kebob" model is composed of a line of beads. **c** The slender-body model represents the rod as a line of Stokeslets. **d** An exact solution for an ellipsoidal particle is another option

resistance and the beads are typically assumed to interact only through the constraint, not through the fluid disturbances. This dumbbell model has been used for a number of simulations of the collective dynamics of particles and is a particularly popular choice for modeling small, Brownian rods (Bird et al. 1971; Bitsanis et al. 1988, 1990).

The so-called "shish-kebob" model (Riseman and Kirkwood 1950), like the dumbbell model, utilizes beads connected by rigid constraints, though the beads are distributed along the entire length of the rod as shown in Fig. 8 instead of merely at the ends of the rod. As one would expect, this model gives improved fidelity for the calculation, though again the beads are assumed to interact only through the constraint, generally.

Accurate solutions for the motion of a single ellipsoidal particle are readily available as well (Oberbeck 1876; Jeffery 1922). The singularity solution of Chwang and Wu (1975) for an ellipsoidal particle has been used as the basis for simulations of the collective dynamics of ellipsoids (Claeys and Brady 1993). The calculation is analogous to the well-known Stokesian dynamics method for simulating the collective dynamics of spheres (Brady and Bossis 1988), but the algorithm is more complex to code and is expensive to implement owing to the change of geometry.

Another option for modeling rods utilizes the hydrodynamic theory of slender bodies (Cox 1970; Batchelor 1970). For any particle of sufficiently high aspect ratio, slender body theory can be used to approximate the hydrodynamic forces. To do so, the slender body is represented by a continuous line of hydrodynamic resistance, where the resistance at each point along the line is chosen to approximate the no-slip condition at the surface. If the cross section is non-uniform or the slender body is curved, the equation is still applicable. For the case of a rigid, straight rod, the equations simplify to a form that is easier to handle than that of the exact solution for an ellipse. Of course, one is giving up accuracy by using the approximation, but in fact the equations are accurate for very high aspect ratio and the equations produce the correct, qualitative behavior, unlike some of the other, simpler models.

4.2 *Slender Body Equation for a Rigid Rod*

The slender body equation relates the line force density, $\boldsymbol{f}$, acting at any point $\boldsymbol{r}$ along the length of the body to the difference between the velocity of the rod and fluid at that point,

$$\boldsymbol{f}(\boldsymbol{r}) = \frac{4\pi\mu}{\ln(2A)} \left(\boldsymbol{I} - \frac{1}{2}\boldsymbol{p}(\boldsymbol{r})\boldsymbol{p}(\boldsymbol{r}) \right) \cdot \left(\dot{\boldsymbol{r}}(\boldsymbol{r}) - \boldsymbol{u}^{\infty}(\boldsymbol{r}) \right). \tag{21}$$

The ratio of the total length of the body, L, and the diameter d is the aspect ratio, A, the viscosity is μ, and the unit vector $\boldsymbol{p}(\boldsymbol{r})$ is tangent to the body at each point $(\boldsymbol{r})$. The line force density is proportional to the time rate of change of the position of the line, $\dot{\boldsymbol{r}}(\boldsymbol{r}) = \partial \boldsymbol{r}/\partial t$, and the velocity of the fluid at the same point, $\boldsymbol{u}^{\infty}(\boldsymbol{r})$. These variables and the geometry are diagrammed in Fig. 9 for clarity.

The right hand side of Eq. 21 is the leading order calculation for the hydrodynamic force acting on the particle; higher order corrections, along with detailed derivations, can be found in Batchelor (1970) and Cox (1970). The left side is the sum of the

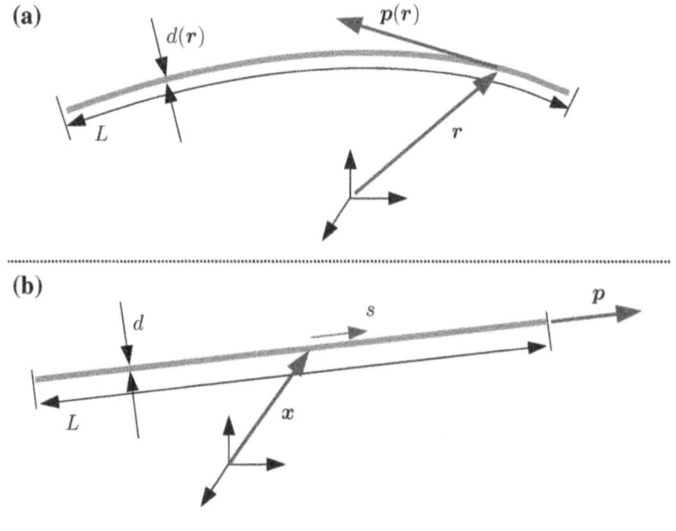

Fig. 9 The geometry and variables describing a slender body. **a** For a general slender body, the diameter $d(\boldsymbol{r})$ can be a function of position $\boldsymbol{r}$ along the length L, so long as the diameter at every point is much smaller than the length. The tangent to the line at each point is $\boldsymbol{p}(\boldsymbol{r})$. **b** The rigid rod has a constant diameter d, where the aspect ratio $A = L/d$ is much larger than one. The position at any point of the centerline of the rod is given by $\boldsymbol{x} + s\boldsymbol{p}$, where s is measured from the center as shown

forces acting on the rod, save for the hydrodynamic force; note that $\boldsymbol{f}$ is a force per unit length, rather than traction (force per unit area) or simply a force. The line force density approximates the normal component of the surface traction (stress) integrated over the perimeter of the body, at the point $\boldsymbol{r}$.

Here, the interest is in a slender body that is straight and rigid. In this case, the position $\boldsymbol{r}$ can be parameterized as $\boldsymbol{x} + s\boldsymbol{p}$, where the center of the rod is located at $\boldsymbol{x}$, the unit vector $\boldsymbol{p}$ is no longer a function of position along a rod, and s is a local coordinate on the rod that ranges from $-L/2$ to $L/2$. These coordinates are diagrammed in Fig. 9.

To solve for the motion of a rigid rod, Eq. 21 is solved to give an explicit expression for $\dot{\boldsymbol{r}}(\boldsymbol{r})$, and then use is made of the constraint that $\boldsymbol{r} = \boldsymbol{x} + s\boldsymbol{p}$:

$$\dot{\boldsymbol{x}} + s\dot{\boldsymbol{p}} = \frac{\ln(2A)}{4\pi\mu}\left(\boldsymbol{I} + \boldsymbol{pp}\right) \cdot \boldsymbol{f}(\boldsymbol{x} + s\boldsymbol{p}) + \boldsymbol{u}^{\infty}(\boldsymbol{x} + s\boldsymbol{p}). \tag{22}$$

Note that to invert Eq. 21, the identity,

$$(\boldsymbol{I} + c\boldsymbol{pp}) \cdot \left(\boldsymbol{I} - \frac{c}{1+c}\boldsymbol{pp}\right) = \boldsymbol{I}, \tag{23}$$

must be utilized, where c is an arbitrary constant that can not equal -1.

Typically when solving a mobility or resistance problem, the moments of the force (total force, torque, stresslet, and even higher terms) are the quantities which should be related to the motion. To generate a simulation method analogous to Stokesian dynamics, the effect of the total force, torque, and stresslet must be calculated.

Integrating Eq. 22 over the length of the rod demonstrates that a force on the rod results in a center of mass motion,

$$\dot{\boldsymbol{x}} = \frac{\ln(2A)}{4\pi\mu L}\left(\boldsymbol{I} + \boldsymbol{pp}\right) \cdot \boldsymbol{F} + \frac{1}{L}\int_{-L/2}^{L/2} \boldsymbol{u}^{\infty}(s)ds, \tag{24}$$

where the integral over $s = -L/2$ to $L/2$ should be understood to range over $\boldsymbol{x} - L\boldsymbol{p}/2$ to $\boldsymbol{x} + L\boldsymbol{p}/2$. The force acting on the rod is given by

$$\boldsymbol{F} = \int_{-L/2}^{L/2} \boldsymbol{f}(s)ds, \tag{25}$$

where $\boldsymbol{f}(s)$ is the sum of the forces acting at each point on the rod.

Likewise, integrating over the length of the rod after taking the cross product of $s(\boldsymbol{p} \times \boldsymbol{p})$ with Eq. 22 gives,

$$\dot{\boldsymbol{p}} = -\frac{3\ln(2A)}{\pi\mu L^3}\boldsymbol{p} \times \boldsymbol{\tau} + \frac{12}{L^3}\left(\boldsymbol{I} - \boldsymbol{pp}\right) \cdot \int_{-L/2}^{L/2} s\boldsymbol{u}^{\infty}(s)ds, \tag{26}$$

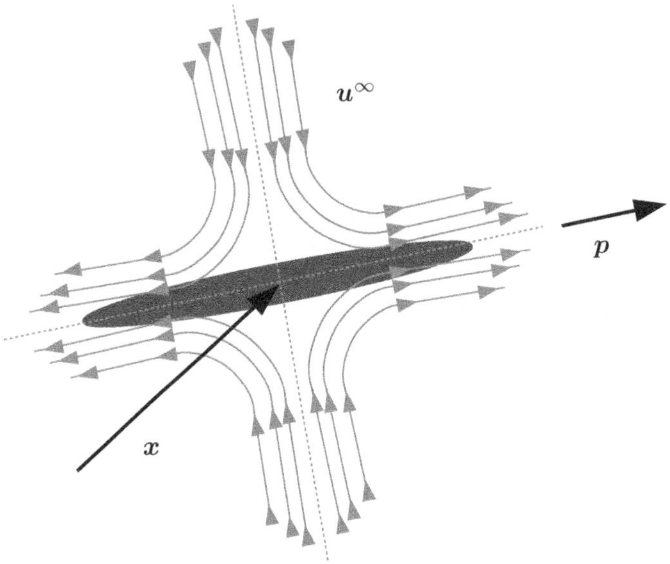

Fig. 10 Straining components of the imposed flow aligned with the rod results in no stretching (or compressing) of the rigid body, nor any other motion of the rod. However, the straining flow does contribute to the hydrodynamic force on the rod, and hence generates a disturbance to the velocity of the fluid

where

$$\tau = \int_{-L/2}^{L/2} \boldsymbol{p} \times s\boldsymbol{f}(s)ds. \tag{27}$$

In deriving Eq. 26, the vector identity

$$\boldsymbol{p} \times (\boldsymbol{p} \times \boldsymbol{v}) = -(\boldsymbol{I} - \boldsymbol{pp}) \cdot \boldsymbol{v} \tag{28}$$

is utilized, where $\boldsymbol{v}$ is an arbitrary vector. Also, the change in orientation with time, $\dot{\boldsymbol{p}} = \partial \boldsymbol{p}/\partial t$, can have no components parallel to the orientation since the rod is rigid. Hence, the constraint that

$$(\boldsymbol{I} - \boldsymbol{pp}) \cdot \dot{\boldsymbol{p}} = \dot{\boldsymbol{p}}, \tag{29}$$

or $\boldsymbol{p} \cdot \dot{\boldsymbol{p}} = 0$, is employed when writing the final form of Eq. 26 for the rotational velocity $\dot{\boldsymbol{p}}$.

Though the component of straining flow aligned with the rod does not generate a motion of the particle, it does generate a stress on the rod and, hence, a hydrodynamic force (see Fig. 10). To compute this component, Eq. 22 is again integrated over the length of the rod, but the inner product of both sides of the equation is made first with $s\boldsymbol{p}$. The particle velocities disappear, leaving an equality between the line force density and velocity of the fluid,

$$S = -\frac{2\pi\mu}{\ln(2A)} \int_{-L/2}^{L/2} \left[s\boldsymbol{p} \cdot \boldsymbol{u}^\infty \right] ds = \int_{-L/2}^{L/2} \left[s\boldsymbol{p} \cdot \boldsymbol{f}(s) \right] ds. \tag{30}$$

Higher order moments of the hydrodynamic forces can be calculated in a similar manner. However, for the purposes of simulating the collective motion of a concentrated suspension of rods, the linear components of the force distribution are sufficient, as will be argued later in this document.

4.3 Motion of a Slender Rod

Three examples are briefly given to demonstrate key features of the motion of rods that can be expected from solutions of the slender-body approximation. These solutions are compared and contrasted with the expected results as well as the motions of a sphere under similar circumstances.

In the first example, the motion of a rod falling through a quiescent fluid ($\boldsymbol{u}^{\infty} = 0$) due to a body force such as gravity is examined. The rod is assumed to have a uniform density, so that the net torque on the rod is zero and it does not rotate as it falls under these conditions of negligible inertia. The center of mass motion due to the gravitational force $\boldsymbol{F}_g$ (which must include buoyancy forces) is given by Eq. 24,

$$\dot{\boldsymbol{x}} = \frac{\ln(2A)}{4\pi\mu L}(\boldsymbol{I} + \boldsymbol{pp}) \cdot \boldsymbol{F}_g \tag{31}$$

or equivalently

$$\dot{\boldsymbol{x}} = \frac{\ln(2A)}{4\pi\mu L}\left(\boldsymbol{F}_g + \boldsymbol{p}\left(\boldsymbol{p} \cdot \boldsymbol{F}_g\right)\right). \tag{32}$$

If the fiber is aligned with gravity or perpendicular to gravity, the velocity is solely in that direction (see Fig. 11a). For alignment with gravity, the slender body model predicts $U_{\|} = 2F_g \ln(2A)/4\pi\mu L$, which is twice the velocity $U_{\perp}$ of a fiber that is aligned perpendicular to gravity. This factor of two difference represents an unreal-

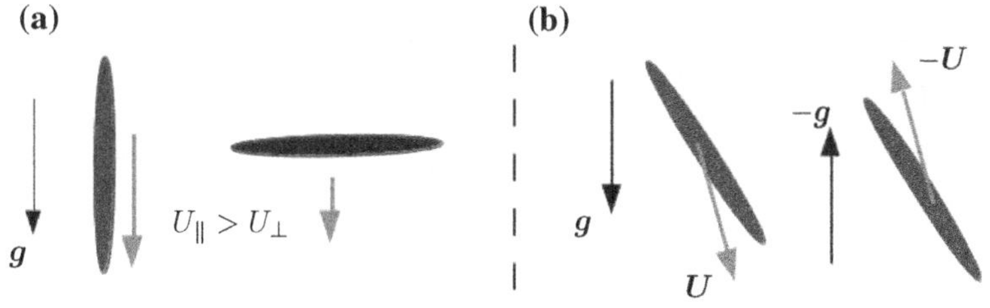

Fig. 11 The properties of a single sedimenting rod. **a** When aligned with gravity, the sedimentation velocity ($U_{\|}$) exceeds that of the sedimentation velocity when aligned perpendicular to gravity ($U_{\perp}$) by up to a factor of two. **b** When aligned at an angle with respect to gravity, the rod velocity will have components of motion that are transverse to the gravitational force; reversing the direction of the force (gravity) must result in a reversal of the velocity

istic, theoretical limit for the ratio of the parallel and perpendicular velocities that is a result of the slender body model. The velocity $U_{\parallel}$ should exceed $U_{\perp}$, but predicting the correct ratio depends upon a more quantitative model than used here, or at least depends upon adding corrections to the leading order slender body model.

Still, the leading order calculations from the slender body model provide a good, qualitative description of the particle motion. For example, Eq. 32 correctly predicts that a rod sedimenting at an angle with respect to gravity (see Fig. 11b) will have components of motion that are perpendicular to gravity. This is very different from the sedimentation of a sphere, where there can be no components of motion perpendicular to the applied force for a single particle. The motion of even individual rods consequently can exhibit dynamic trajectories that are significantly more complex than single spheres. For example, consider the motion of a single particle falling between two bounding walls as illustrated in Fig. 12. A spherical particle will fall straight through the channel, though the presence of the walls will cause the sphere to rotate with a rate that is proportional to its lateral position. The rod, however, will drift in the channel since the presence of the bounding walls causes the rod to rotate and then drift side-to-side as it falls (Russel et al. 1977). Importantly, the motion of the rods for the cases shown in Figs. 11 and 12 is reversible, as required for the motion of a rod in a viscous fluid in the limit of zero Reynolds number.

As a second example, the motion of a force and torque-free rod in a simple shear flow is calculated. For a shear-flow of rate $\dot{\gamma}$ with gradient in the y-direction and flow in the x-direction,

$$\boldsymbol{u}^{\infty}(\boldsymbol{x}) = \dot{\gamma} y \boldsymbol{\delta}_x, \tag{33}$$

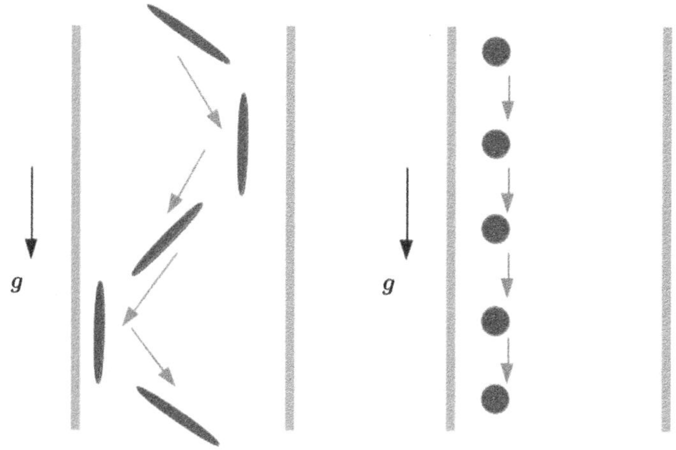

Fig. 12 Sedimentation of a particle between bounding walls. Interactions between the walls and the sedimenting particle cause the particles to rotate, unlike the case in an unbounded fluid. For the rod, the rotation couples with the drift motion (see Fig. 11) to generate the side-to-side motion shown on the *left*. For the spherical particle (on *right*), the rotation is not coupled with the center-of-mass motion and the particle falls straight through the channel

the center-of-mass of the rod moves with the velocity as evaluated at its center,

$$\dot{\boldsymbol{x}} = \dot{\gamma} x_y \boldsymbol{\delta}_x, \tag{34}$$

where x_y is the y-component of the center position of the rod. The rotation of the rod reduces to

$$\dot{\boldsymbol{p}} = \frac{12}{L^3} (\boldsymbol{I} - \boldsymbol{pp}) \cdot \int_{-L/2}^{L/2} s\dot{\gamma} \left(x_y + s p_y\right) \boldsymbol{\delta}_x ds \tag{35}$$

after substituting Eq. 33 into 26 and setting $\boldsymbol{\tau}$ to zero. Computing the integral and using $(\boldsymbol{I} - \boldsymbol{pp}) \cdot \boldsymbol{\delta}_x = \boldsymbol{\delta}_x - p_x \boldsymbol{p}$ gives

$$\dot{\boldsymbol{p}} = \dot{\gamma} \left(p_y \boldsymbol{\delta}_x - p_x p_y \boldsymbol{p}\right). \tag{36}$$

The model predicts that the rod orientation does not change if the rod is aligned in the flow-vorticity (x-z) plane ($p_y = 0$). In fact, rods aligned in this manner will rotate through the flow-gradient plane, even as the aspect ratio becomes large. This is another consequence of the leading order slender-body approximation that can be corrected using the theoretical work of Jeffery (1922).

As with sedimentation, adding walls demonstrates that the dynamics for the shearing flow of rod-like particles is much richer than the dynamics of spherical particles, at least in the limit of zero Reynolds. Figure 13 illustrates the motion of a sphere in a shear flow parallel to a bounding wall. For a viscous flow in the absence of inertia, the sphere moves parallel to the wall while also rotating. In the case of a rod, the rod can move across streamlines, where the direction and rate of migration depends upon the orientation (Hsu and Ganatos 1976; Park et al. 2007; Park and Butler 2009). This motion is not a violation of the principle that particle motion must be reversible; as shown in Fig. 13, the reversal of the flow should lead to a reversal of the migration direction and does according to the calculations.

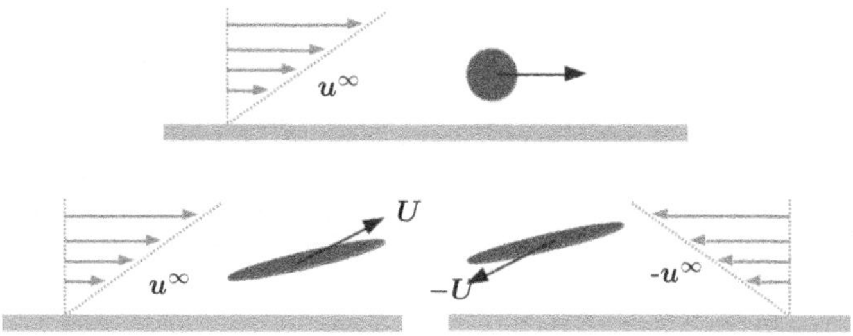

Fig. 13 The motion of a particle in a shear flow near to a bounding wall. In the limit of zero Reynolds number, a sphere in a shearing flow near a bounding wall (*top*) will not migrate across streamlines. However, a rod under similar conditions can migrate across streamlines with a velocity that depends upon its instantaneous orientation. The motion is reversible, as shown

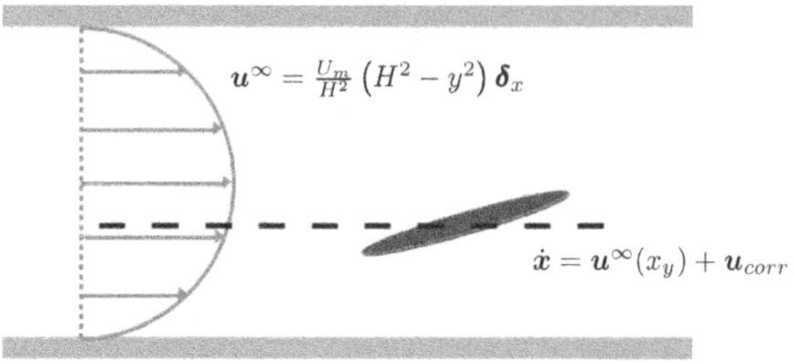

Fig. 14 As with spheres, the motion of a rod contains a correction to the velocity at the center of the particle for flows that are not linear. For the parabolic flow shown here, the rod moves with the velocity of the fluid at its center, $\boldsymbol{u}^\infty(x_y)$, and a correction $\boldsymbol{u}_{corr}$ due to the curvature of the flow

In linear flows, the center-of-mass motion of a rod matches the velocity of the fluid flow as evaluated at its center-of-mass; the same is true for spheres, though it is well known that there is a correction, referred to as Faxen's Law, for the motion of spheres within flows that are not linear. In this third example of particle motion, the motion of a rod in a non-linear flow is examined. As shown in Fig. 14, the flow field considered is a pressure-driven flow between two planar walls separated by a distance H with maximum velocity of U_m,

$$\boldsymbol{u}^\infty = \frac{U_m}{H^2}\left(H^2 - y^2\right)\boldsymbol{\delta}_x, \tag{37}$$

where the maximum is at the center of the channel ($y = 0$) and the flow is in the x-direction. Calculating the center of mass velocity of a rod, using Eq. 24 and the velocity field above, gives

$$\dot{\boldsymbol{x}} = \frac{U_m}{H^2}\left(H^2 - x_y^2\right)\boldsymbol{\delta}_x - \frac{U_m L^2}{12H^2}p_y\boldsymbol{\delta}_x. \tag{38}$$

The first term is the motion of the rod which corresponds to the velocity of the fluid as evaluated at the center of the rod (i.e. $\boldsymbol{u}^\infty(x_y)$). The second term represents the correction to the rod velocity due to the curvature of the flow field. The correction slows the rod, relative to the velocity at its center, by an amount proportional to the orientation component p_y; the correction is less important as the particle length L decreases relative to the wall spacing H and is purely in the direction of the flow, as the bounding walls are not considered in the analysis.

5 Simulating Rigid Rods

The slender body approximation can be used as the basis of simulating the collective motion of rods, in a manner similar to the simulation of spheres using the well-known Stokesian dynamics method (Brady and Bossis 1988). As with Stokesian dynamics,

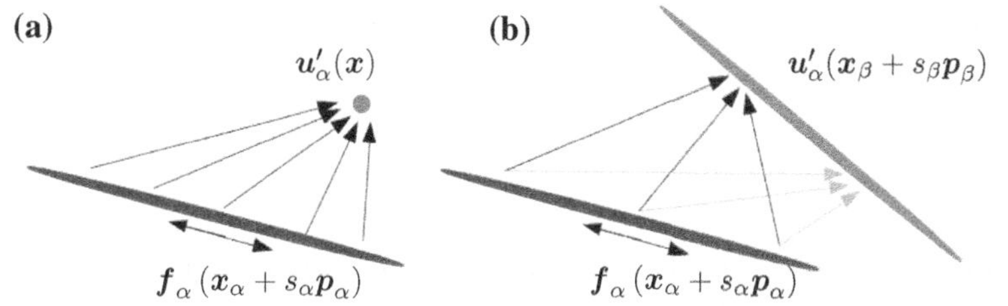

Fig. 15 **a** The disturbance caused by a rod α can be calculated at any point $\boldsymbol{x}$ by integrating over the length of the rod. **b** To calculate the motion of a rod β in the presence of the rod α, the velocity disturbance caused by rod α can be evaluated at every point along the centerline of β and then integrated to give the motion

the method for simulating rods should incorporate the multi-body hydrodynamics and provide a means for evaluating the rheology. Here, such a method is described.

5.1 *Hydrodynamic Interactions*

For the purpose of calculating the collective dynamics of a group of rods, the influence of the motion of one particle on another must be included. If the rods are non-colloidal and the possibility of direct solid contacts is discounted, then interactions arise from the disturbance velocity caused by the hydrodynamic forces on a rod. For example, a point force acting on a fluid generates a motion given by Eq. 7 and as shown in Fig. 5. Recognizing that the slender-body rod is a continuous line of forces, the disturbance velocity due to any rod α can be calculated at any point $\boldsymbol{x}$ by summing (integrating) the disturbances created by each point over the length of the fiber,

$$\boldsymbol{u}'_\alpha(\boldsymbol{x}) = \int_{-L/2}^{L/2} \boldsymbol{G}\left(\boldsymbol{x} - (\boldsymbol{x}_\alpha + s\boldsymbol{p}_\alpha)\right) \cdot \boldsymbol{f}(\boldsymbol{x}_\alpha + s\boldsymbol{p}_\alpha) ds, \tag{39}$$

where $\boldsymbol{x}_\alpha$ and $\boldsymbol{p}_\alpha$ are the center of mass and orientation of rod α. Furthermore as illustrated in Fig. 15, the velocity can be calculated at any point on another fiber β.

To calculate the motion of a rod β which incorporates the hydrodynamic influence of surrounding rods, Eqs. 24 and 26 can be altered,

$$\dot{\boldsymbol{x}}_\beta = \frac{\ln(2A)}{4\pi\mu L} \left(\boldsymbol{I} + \boldsymbol{p}_\beta \boldsymbol{p}_\beta\right) \cdot \boldsymbol{F}_\beta + \frac{1}{L} \int_{s_\beta=-L/2}^{L/2} \boldsymbol{u}(s_\beta) ds_\beta \tag{40}$$

$$\dot{\boldsymbol{p}}_\beta = -\frac{3\ln(2A)}{\pi\mu L^3} \boldsymbol{p}_\beta \times \boldsymbol{\tau}_\beta + \frac{12}{L^3} \left(\boldsymbol{I} - \boldsymbol{p}_\beta \boldsymbol{p}_\beta\right) \cdot \int_{s_\beta=-L/2}^{L/2} s_\beta \boldsymbol{u}(s_\beta) ds_\beta, \tag{41}$$

where $\boldsymbol{u}(s_\beta)$ has been substituted in place of $\boldsymbol{u}^\infty(s_\beta)$. The velocity represents the sum of the background velocity, $\boldsymbol{u}^\infty(s_\beta)$, and the disturbance velocity caused by all other particles,

$$\boldsymbol{u}(s_\beta) = \boldsymbol{u}^\infty(s_\beta) + \sum_\alpha \boldsymbol{u}'_\alpha(s_\beta); \tag{42}$$

the summation over the disturbance velocities should exclude the disturbance velocity of the rod β itself. Note that the velocities add since the fluid motion is governed by Stokes equation, which is linear, unlike the case of a fluid motion that is governed by the Navier–Stokes equation where finite inertia enters into the calculation.

5.2 *Solving for the Collective Dynamics*

Equations 39–42 provide a straightforward method of calculating the collective motion of a group of rods that are widely separated and suspended in an unbounded, viscous fluid. Equations 40 and 41 are arranged so as to evaluate the velocities explicitly (the mobility problem) given the forces, but the equations can be inverted to solve for the forces (the resistance problem) if the velocities are known. Likewise, the equations can be arranged to solve so-called 'mixed problems', wherein a mixture of velocities and forces are to be calculated from a combination of specified velocities and forces. Also, the method can be utilized to solve for the motion in a variety of geometries other than an unbounded fluid by replacing the Oseen tensor in Eq. 7 with the appropriate Green's function. Readily available Green's functions include those for periodic geometries (Hasimoto 1959), for infinite plane walls (Blake 1971), and Pozrikidis (1992) lists many more.

An important point is that even if the total force and torque are given on each rod, the line force density is generally unknown and must be calculated as part of the solution procedure. One approach is to discretize the length of each rod and solve the set of equations for the force distribution and the velocity disturbance on each rod (Mackaplow and Shaqfeh 1998). An alternative is to expand the velocity disturbance along the length of a fiber in a Legendre polynomial and then to solve for the line-force density (Harlen et al. 1999; Butler and Shaqfeh 2002). Retaining only the linear terms gives the approximation

$$\boldsymbol{f}(s) \approx \frac{1}{L}\boldsymbol{F} + \frac{12s}{L^3}\left(\boldsymbol{\tau} \times \boldsymbol{p} + S\boldsymbol{p}\right). \tag{43}$$

The resulting set of equations for the motion of multiple particles is a multibody problem which formally requires consideration of an infinite number of calculations of the induced disturbances, much like other problems in physics which involve interactions between multiple objects such as the n-body problem in celestial mechanics (Aarseth 2003) and the many-body problem in quantum mechanics (Hunziker and Sigal 2000). Consider the motion of only three rods, each acted upon by a force:

the disturbance generated by the force on each rod results in a disturbance on the other two. The presence of the disturbance on any one rod due to the other two alters the force distribution, and this alteration of the force distribution must be considered when calculating the disturbance velocity. This cycle of reflecting the velocity disturbances and force distributions continues endlessly and would appear to be insolvable.

As with all such multi-body problems, simulating the collective dynamics of the rods requires approximations, and a very convenient one is available. Each reflection of an interaction is significantly weaker than the previous, by a factor of

$$\frac{1}{\ln(2A)}\frac{L}{r}, \tag{44}$$

where r is a measure of the separation distance between the rods. For M reflections, the change to the motion of a rod would be

$$\left(\frac{1}{\ln(2A)}\frac{L}{r}\right)^M. \tag{45}$$

For widely spaced rods of high aspect ratio, the error in truncating the interactions after $M = 2$ or 3 is small. Such an approach is commonly used when simulating the collective motion of spheres where the reflections are stronger. For spheres of radius a, each reflection M of the interactions contributes a velocity that scales as a fraction

$$\left(\frac{a}{r}\right)^M \tag{46}$$

of the leading contribution to the velocity (Kim and Karrila 2005).

Note that the rapid decay of the interactions given by Eq. 44 is the basis of the frequent claim that hydrodynamic interactions can be ignored in suspensions of slender bodies. While true for relatively dilute suspensions (i.e. L/r small) for rods as the aspect ratio goes to infinity, the approximation must be used with caution. Even in this limit of dilute concentrations, fluctuations in the concentration that generate pairs of nearby rods can, for example, make the approximation invalid.

5.3 Example Calculation

A proof is not offered for Eq. 44, but rather a demonstrative calculation is given here using a method of reflections. The example will calculate the motion of a rod sedimenting through a quiescent, viscous fluid in the vicinity of a second rod. The motion of the rod of interest (α) is given by

$$\dot{\boldsymbol{x}}_\alpha = \frac{\ln(2A)}{4\pi\mu L}\left(\boldsymbol{I}+\boldsymbol{p}_\alpha\boldsymbol{p}_\alpha\right)\cdot\boldsymbol{F}_\alpha, \tag{47}$$

where $\boldsymbol{F}_\alpha$ is the net force on the rod. The first correction to the motion of rod α is due to the velocity disturbance caused by the force $\boldsymbol{F}_\beta$ acting on the second rod,

$$\boldsymbol{u}'_\beta(\boldsymbol{x},\boldsymbol{F}_\beta) = \frac{1}{L}\int_{-L/2}^{L/2}\boldsymbol{G}\left(\boldsymbol{x}-(\boldsymbol{x}_\beta+s_\beta\boldsymbol{p}_\beta)\right)\cdot\boldsymbol{F}_\beta ds_\beta. \tag{48}$$

This velocity disturbance should be included within the motion of rod α by evaluating the disturbance at each point along rod α and integrating along the length, as indicated in Eq. 24,

$$\frac{1}{L}\int_{-L/2}^{L/2}\boldsymbol{u}'_\beta(\boldsymbol{x},\boldsymbol{F}_\beta)ds_\alpha = \frac{1}{L^2}\int_{-L/2}^{L/2}\int_{-L/2}^{L/2}\boldsymbol{G}\left((\boldsymbol{x}_\alpha+s_\alpha\boldsymbol{p}_\alpha)-(\boldsymbol{x}_\beta+s_\beta\boldsymbol{p}_\beta)\right)ds_\beta ds_\alpha\cdot\boldsymbol{F}_\beta \tag{49}$$

This correction is added to Eq. 48 to give an improved calculation of the motion of the rod.

Likewise the force on rod α generates a velocity disturbance, $\boldsymbol{u}'_\alpha(\boldsymbol{x},\boldsymbol{F}_\alpha)$, that alters the sedimentation velocity of the rod β; the calculation is similar to that given above. Since there is now a (disturbance) flow on rod β, there is an additional (hydrodynamic) force acting on the rod which generates a stresslet S and may also cause the rod to rotate. The force moment of interest is calculated from Eq. 30,

$$S_\beta = -\frac{2\pi\mu}{\ln(2A)}\int s_\beta\boldsymbol{p}_\beta\cdot\boldsymbol{u}'_\alpha((\boldsymbol{x}_\alpha+s_\alpha\boldsymbol{p}_\alpha)-(\boldsymbol{x}_\beta+s_\beta\boldsymbol{p}_\beta),\boldsymbol{F}_\alpha)ds_\beta, \tag{50}$$

where the velocity disturbance can be calculated and inserted to give

$$S_\beta = -\frac{2\pi\mu}{L\ln(2A)}\boldsymbol{p}_\beta\cdot\int\int\boldsymbol{G}\left((\boldsymbol{x}_\alpha+s_\alpha\boldsymbol{p}_\alpha)-(\boldsymbol{x}_\beta+s_\beta\boldsymbol{p}_\beta)\right)ds_\alpha s_\beta ds_\beta\cdot\boldsymbol{F}_\alpha. \tag{51}$$

This moment of the force distribution on rod β creates the next correction, or reflection, that is added to the motion of the rod of interest, α. The velocity disturbance created by the inability of the rod β to elongate further alters the flow field experienced at every point $\boldsymbol{x}_\alpha+s_\alpha\boldsymbol{p}_\alpha$ by an amount of

$$\boldsymbol{u}'_\alpha((\boldsymbol{x}_\alpha+s_\alpha\boldsymbol{p}_\alpha)-(\boldsymbol{x}_\beta+s_\beta\boldsymbol{p}_\beta),S_\beta) = \int\boldsymbol{G}\left((\boldsymbol{x}_\alpha+s_\alpha\boldsymbol{p}_\alpha)-(\boldsymbol{x}_\beta+s_\beta\boldsymbol{p}_\beta)\right)\left(\frac{12s_\beta}{L^3}S_\beta\boldsymbol{p}_\beta\right)ds_\beta. \tag{52}$$

This velocity is integrated along the rod to give the additional contribution of

$$\dot{\boldsymbol{x}}_\alpha = \cdots + \frac{12S_\beta}{L^4} \int\int \boldsymbol{G}\left((\boldsymbol{x}_\alpha + s_\alpha \boldsymbol{p}_\alpha) - (\boldsymbol{x}_\beta + s_\beta \boldsymbol{p}_\beta)\right) \cdot s_\beta \boldsymbol{p}_\beta ds_\beta ds_\alpha. \tag{53}$$

Each contribution to the velocity originates with the forces $\boldsymbol{F}_\alpha$ and $\boldsymbol{F}_\beta$ that act on the two rods. Assuming that the two forces have an equivalent magnitude of F indicates that the leading order calculation of the velocity is given by

$$\frac{F \ln(2A)}{4\pi\mu L}, \tag{54}$$

as seen in Eq. 47. Comparing this with the next contribution to the velocity of the rod (Eq. 49) indicates that this term is smaller by a factor of

$$\frac{1}{\ln(2A)} \frac{L}{r}, \tag{55}$$

since the contribution of the Oseen tensor ($\boldsymbol{G}$, Eq. 7) scales as the inverse separation distance of the particles, which is indicated by r. The subsequent contribution of Eq. 53 is smaller again by the same factor, as is the contribution to the motion of rod α from each additional reflection of the hydrodynamic interactions. Continuing the calculation as above would demonstrate the more general result that, after M reflections, the change to the velocity of the rod is given by Eq. 45.

5.4 Slender Body Dynamics

To compute the motion of multiple rods, the equations for the motion and the interactions can be combined with the approximation of the line force distribution on each rod to give a set of equations for the collective motion of the rods,

$$\begin{bmatrix} \dot{\boldsymbol{X}} - \dot{\boldsymbol{X}}^\infty \\ \dot{\boldsymbol{P}} - \dot{\boldsymbol{P}}^\infty \\ -\boldsymbol{U}_s^\infty \end{bmatrix} = \mathcal{M} \cdot \begin{bmatrix} \mathcal{F} \\ \mathcal{T} \\ \mathcal{S} \end{bmatrix}. \tag{56}$$

In these equations, the sub-vector $\dot{\boldsymbol{X}}$ contains the center of mass motions for each rod in each direction and, for a system of two rods, would be written as

$$\dot{\boldsymbol{X}} = \begin{bmatrix} \dot{x}_1 \\ \dot{y}_1 \\ \dot{z}_1 \\ \dot{x}_2 \\ \dot{y}_2 \\ \dot{z}_2 \end{bmatrix}, \tag{57}$$

where $\dot{x}$, $\dot{y}$, and $\dot{z}$ indicate the velocities in each direction and the subscript indicates the particle number. The expression can be generalized easily for more particles. Likewise, the vector $\dot{\boldsymbol{P}}$ contains the rotational velocities, and $\mathcal{F}$ and $\mathcal{T}$ contain the corresponding forces and torques on the particles.

The sub-vectors $\dot{\boldsymbol{X}}^{\infty}$ and $\dot{\boldsymbol{P}}^{\infty}$ in Eq. 56 contain the motions due to the imposed fluid. For example, the fourth entry in $\dot{\boldsymbol{X}}^{\infty}$ is

$$\frac{1}{L}\int_{-L/2}^{L/2} u_x^{\infty}\left(\boldsymbol{x}_2 + s\boldsymbol{p}_2\right) ds \tag{58}$$

and the fourth entry in $\dot{\boldsymbol{P}}^{\infty}$ is

$$\frac{12}{L^3}\left(\int_{-L/2}^{L/2} s u_x^{\infty}\left(\boldsymbol{x}_2 + s\boldsymbol{p}_2\right) ds - p_{x,2}\boldsymbol{p}_2 \cdot \int_{-L/2}^{L/2} s\boldsymbol{u}^{\infty}\left(\boldsymbol{x}_2 + s\boldsymbol{p}_2\right) ds\right), \tag{59}$$

where u_x^{∞} is the velocity in the x-direction and $p_{x,2}$ is the x-component of the orientation for the second particle.

Unlike the other terms, there is only one entry for each particle in the sub-vectors $\boldsymbol{U}_s^{\infty}$ and $\mathcal{S}$, as seen in Eq. 30. For two particles, the subvectors are

$$\boldsymbol{U}_s^{\infty} = \begin{bmatrix} -\frac{2\pi\mu}{\ln(2A)}\int_{-L/2}^{L/2}\left[s\boldsymbol{p}_1 \cdot \boldsymbol{u}^{\infty}\left(\boldsymbol{x}_1 + s\boldsymbol{p}_1\right)\right] ds \\ -\frac{2\pi\mu}{\ln(2A)}\int_{-L/2}^{L/2}\left[s\boldsymbol{p}_2 \cdot \boldsymbol{u}^{\infty}\left(\boldsymbol{x}_2 + s\boldsymbol{p}_2\right)\right] ds \end{bmatrix} \tag{60}$$

and

$$\mathcal{S} = \begin{bmatrix} S_1 \\ S_2 \end{bmatrix} \tag{61}$$

The mobility matrix $\mathcal{M}$ in Eq. 56 is of size $7N \times 7N$, where N is the number of rods. The blocks along the diagonal contain the self mobilities, which give the motion of each rod due to the forces and torques. The off-diagonal blocks contain the couplings between the moments of the line-force density on any rod to the other rods in the suspension. Each of the entries in $\mathcal{M}$ depends only on the configuration (position and orientation) of the multiple particles and are calculated from double integrals over the Green's function, similar to those seen in the example calculation in Sect. 5.3. The integrals can be evaluated using any numerical integration procedure, such as Gaussian quadrature. The resulting matrix is symmetric and positive definite.

5.5 Lubrication

The above discussion and analysis of hydrodynamic interactions between rods relies on the assumption that the rods are widely separated. In concentrated suspensions, the separation distance between particles can drop below a small fraction of the length, or even diameter, of the rods. In this case, calculating the interaction terms can be done by performing reflections as in Sect. 5.3, but only in principal: the number of reflections needed becomes prohibitive as the separation distance between the particles decreases.

Instead, lubrication theory is used to estimate the interaction between closely spaced particles. For suspensions of spheres of identical size, lubrication interactions are relatively easy to include in simulations with high accuracy for two reasons. Firstly, the lubrication interactions depend upon the curvatures of the surfaces and these are constants for spheres. Secondly, the lubrication interactions between two spheres are known exactly (Arp and Mason 1977; Jeffrey and Onishi 1984; Kim and Mifflin 1985), and hence can be incorporated easily into the mobility matrix.

For suspensions of rods, the situation is more problematic. The lubrication interactions are a function of not only the separation distance, but also the relative alignment and position since the curvature varies across the surface of the rods. To calculate the limiting values, the lubrication approximations derived from the formulas of Claeys and Brady (1989) can be used, but the curvatures of both rods must be calculated at the points of closest approach. These analytical corrections are added to Eq. 56. The lubrication approximations are available, most naturally, in terms of a resistance, rather than mobility, formulation. Hence, the mobility matrix, $\mathcal{M}$, is inverted to give the resistance matrix and then each term is added. Inverting again gives the corrected mobility matrix which contains information about the long and short range interactions.

5.6 Solving

Equation 56 can be rearranged as needed to solve any particular set of conditions. For a mobility problem, where the forces, torques, and flow are specified, the vector $\mathcal{S}$ must first be resolved, as it is a function of the imposed flow. As with Stokesian dynamics (Durlofsky, Brady, and Bossis 1987; Brady et al. 1988), solving for $\mathcal{S}$ before resolving the motions $\dot{\boldsymbol{X}}$ and $\dot{\boldsymbol{P}}$ gives a multi-body solutions for the collective motions of the particles. The motions can be solved without incorporating $\mathcal{S}$, but these solutions include only the pair interactions between the particles, at least if lubrication has not been included in the analysis.

Once the velocities are determined, the positions can be updated using a numerical integration. Since the equations of motion are stiff, care must be taken to retain a stable and convergent solution. Hence, when using explicit integration methods, the time step must be very small. The alternative of using a fully implicit method is not practical, as for large numbers of particles the calculation times become prohibitive. Most often, predictor-corrector methods are implemented, such as the Adams–Bashforth method or Runge–Kutta multistep algorithms.

Though not discussed in detail here, one common goal of simulating the collective dynamics of concentrated suspensions is to calculate the rheology. Given the positions, the rheology can be largely calculated using quantities that are already needed for the dynamic simulations. The calculation of the stresses for spheres, from Stokesian dynamics simulations, are described in depth in many papers (Durlofsky, Brady, and Bossis 1987; Brady et al. 1988; Brady and Bossis 1988). For slender bodies, the stresses are closely related to the vector $\boldsymbol{S}$, which is the key component in the stresslet for each rod. Details concerning the calculation of the stresslets can be found in Park and Butler (2009), and the influences of interparticle contacts on the stresses can also be computed (Snook et al. 2014).

6 Collective Dynamics and Chaos

The dynamics of concentrated suspensions of rods, much like spheres, are irreversible under flow. This is despite expectations which are based upon the restrictions that the flows are purely viscous and the particles are non-colloidal.

Figure 16 shows, as one example of irreversible dynamics of the collective motion, the sedimentation of a collection of rods. Using methods described above, the gravitational force that acts upon the rods and the associated velocity disturbances causes the initially well-mixed suspension to demix in time. The instability was predicted by Koch and Shaqfeh (1989) and confirmed numerically (Mackaplow and Shaqfeh 1998; Butler and Shaqfeh 2002). As another example, an initially spherical cloud of rod-like particles settling due to gravity in an infinite fluid exhibits a fascinating

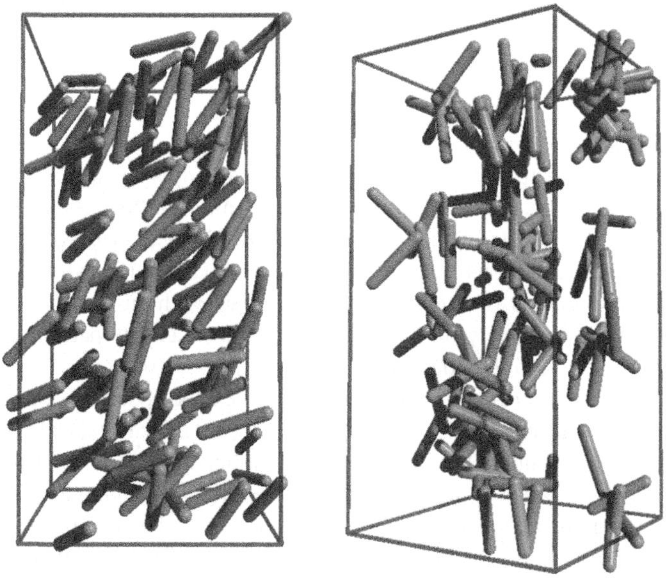

Fig. 16 Initial (*left*) image of a collection of rods in a fully-periodic box is shown. Upon application of a gravitational field, the rods fall through the box and eventually reorganize into clusters of particles, as shown in the image on the *right*

range of dynamic patterns (Park et al. 2010), including the repeated formation and breakup of a toroidal structure. The results are visually similar to those shown in Fig. 7, even though the individual particles are rods instead of spheres.

In both of the above cases, suddenly reversing the direction of the gravitational forces fails to return the particle positions and orientations to their original configuration at time zero. The breaking of the reversibility is, in this case of sedimentation, due to the chaotic motion of rigid rods. For spheres moving due to a net body-force, it is well established (Janosi et al. 1997) that the dynamics are effectively irreversible due to chaos, even though the equations describing the motion are reversible.

For sheared suspensions, of otherwise force-free particles, the question of the origin of irreversibilities is more difficult to discern. Studies (Metzger and Butler 2010; Metzger et al. 2013) indicate that the hydrodynamic interactions in sheared suspensions of spheres are chaotic, but comparisons of models with experimental data (Pham et al. 2015) indicate that additional sources of irreversibility are needed to explain the observations. One prominent possibility is that collisions between rough particles exert an additional displacement. Like suspensions of spheres, studies indicate that the dynamics of sheared suspensions of rods are irreversible and that unique structures can be generated by oscillating the flow (Franceschini et al. 2011). To reproduce the observations from simulations, similar to those described here, requires including short-ranged interparticle forces (Snook et al. 2012); Fig. 17 shows

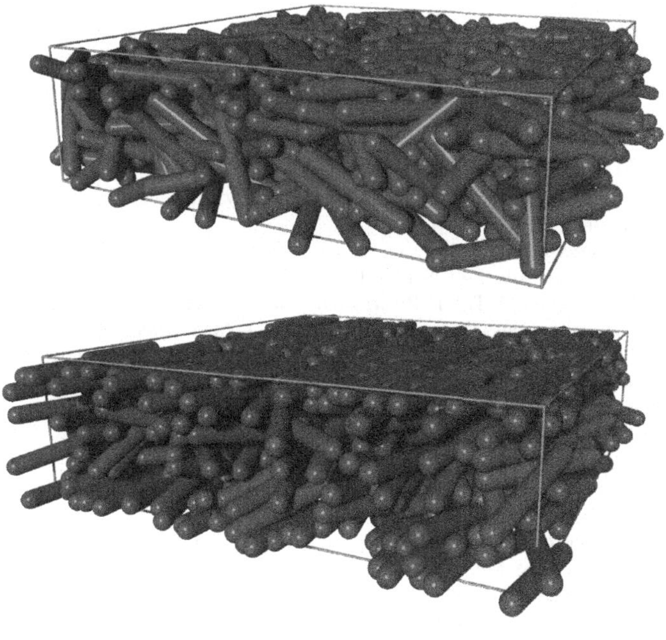

Fig. 17 Initial (*top*) and final (*bottom*) visualizations of an oscillated suspension of rods confined between two bounding walls at high concentration. After multiple oscillations, the rods organize and align with the vorticity direction

an example from the simulations, where the initially random suspension organizes into an aligned one.

7 Concluding Comments

Direct investigations into the chaotic, collective motion of rods are still lacking and need to be performed, though it is increasingly clear that the viscous motion of concentrated, non-colloidal suspensions is irreversible. Methods and analyses, such as those described here, will aid in revealing the key mechanisms that control suspension dynamics and rheology. Also, there are many opportunities for improving simulation capabilities and, perhaps most importantly, using simulations and experimental validations to assist in the generation of macroscopic equations for the accurate prediction of suspension dynamics.

Acknowledgements I thank Mr. Saif Shaikh and Mr. Scott Strednak for their careful reading of this manuscript. The author's work has been supported, in part, by a grant from the National Science Foundation (Grant No. 1511787).

References

S.J. Aarseth. *Gravitational N-Body Simulations*. Cambridge University Press, 2003.

S.A. Altobelli, R.C. Givler, and E. Fukushima. Velocity and concentration measurements of suspensions by nuclear magnetic resonance imaging. J. Rheol., 35: 721–734, 1991.

P.A. Arp and S.G. Mason. The kinetics of flowing dispersions. VIII. Doublets of rigid spheres (theoretical). J. Colloid Interface Sci., 61: 21–43, 1977.

G.K. Batchelor. Slender-body theory for particles of arbitrary cross-section in Stokes flow. J. Fluid Mech., 44: 419–440, 1970.

R.B. Bird, H.R. Warner Jr., and D. C. Evans. Kinetic theory and rheology of dumbbell suspensions with Brownian motion. Adv. Poly. Sci., 8: 1–90, 1971.

I. Bitsanis, H.T. Davis, and M. Tirrell. Brownian dynamics of nondilute solutions of rodlike polymers. 1. Low concentrations. Macromolecules, 21: 2824–2835, 1988.

I. Bitsanis, H.T. Davis, and M. Tirrell. Brownian dynamics of nondilute solutions of rodlike polymers. 2. High concentrations. Macromolecules, 23, 1990.

J.R. Blake. A note on the image system for a Stokeslet in a no-slip boundary. Proc. Cambridge Philos. Soc., 70: 303, 1971.

J.F. Brady and G. Bossis. Stokesian dynamics. Ann. Rev. Fluid Mech., 20: 111–157, 1988.

J.F. Brady, R.J. Phillips, J.C. Lester, and G. Bossis. Dynamic simulation of hydrodynamically interacting suspensions. J. Fluid Mech., 195: 257–280, 1988.

J.E. Butler. Suspension dynamics: moving beyond steady. J. Fluid Mech., 752: 1–4, 2014.

J.E. Butler and R.T. Bonnecaze. Imaging of particle shear migration with electrical impedance tomography. Phys. Fluids, 11: 1982–1994, 1991.

J.E. Butler and E.S.G. Shaqfeh. Dynamic simulations of the inhomogeneous sedimentation of rigid fibres. J. Fluid Mech., 468: 205–237, 2002.

T. Chwang and T.Y. Wu. Hydromechanics of low-Reynolds-number flow. Part 2. Singularity method for Stokes flows. J. Fluid Mech., 67: 787–815, 1975.

I.L. Claeys and J.F. Brady. Lubrication singularities of the grand resistance tensor for two arbitrary particles. Physico Chem. Hydrodyn., 11: 261–293, 1989.
I.L. Claeys and J.F. Brady. Suspensions of prolate spheroids in Stokes flow. Part 1. Dynamics of a finite number of particles in an unbounded fluid. J. Fluid Mech., 251: 411–442, 1993.
R.G. Cox. The motion of long slender bodies in a viscous fluid. Part 1. General theory. J. Fluid Mech., 44: 791–810, 1970.
L. Durlofsky, J.F. Brady, and G. Bossis. Dynamic simulation of hydrodynamically interacting particles. J. Fluid Mech., 180: 21–49, 1987.
A. Franceschini, E. Filippidi, E. Guazzelli, and D. J. Pine. Transverse alignment of fibers in a periodically sheared suspension: An absorbing phase transition with a slowly varying control parameter. Phys. Rev. Lett., 107: 250603, 2011.
E. Guazzelli and J.F. Morris. *A Physical Introduction to Suspension Dynamics*. Cambridge University Press, 2012.
R.E. Hampton, A.A. Mammoli, A.L. Graham, N. Tetlow, and S.A. Altobelli. Migration of particles undergoing pressure-driven flow in a circular conduit. J. Rheol., 41: 621–640, 1997.
O.G. Harlen, R.R. Sundararajakumar, and D.L. Koch. Numerical simulation of a sphere settling through a suspension of neutrally buoyant fibres. J. Fluid Mech., 388: 355–388, 1999.
H. Hasimoto. On the periodic fundamental solutions of the Stokes equations and their application to viscous flow past a cubic array of spheres. J. Fluid Mech., 5: 317–328, 1959.
R. Hsu and P. Ganatos. Gravitational and zero-drag motion of a spheroid adjacent to an inclined plane at low Reynolds number. J. Fluid Mech., 268: 267, 1976.
W. Hunziker and I.M. Sigal. The quantum N-body problem. J. Math. Phys., 41, 2000.
I.M. Janosi, T. Tel, D.E. Wolf, and J.A.C. Gallas. Chaotic particle dynamics in viscous flows: The three-particle Stokeslet problem. Phys. Rev. E, 65: 2858–2868, 1997.
G.B. Jeffery. The motion of ellipsoidal particles immersed in a viscous fluid. Proc. Roy. Soc. A, 102: 161–179, 1922.
J. Jeffrey and Y. Onishi. Calculation of the resistance and mobility functions for two unequal rigid spheres in low-Reynolds-number flow. J. Fluid Mech., 139: 261–290, 1984.
A. Karnis, H. Goldsmith, and S. Mason. The kinetics of flowing dispersions: I. Concentrated suspensions of rigid particles. J. Colloid Interface Sci., 22: 531–553, 1966.
S. Kim and S.J. Karrila. *Microhydrodynamics: Principles and Selected Applications*. Butterworth-Heineman, 2005.
S. Kim and R.T. Mifflin. The resistance and mobility functions of two equal spheres in low-Reynolds-number flow. Phys. Fluids, 28: 2033–2045, 1985.
D.L. Koch and E.S.G. Shaqfeh. The instability of a dispersion of sedimenting spheroids. J. Fluid Mech., 209: 521–542, 1989.
D. Leighton and A. Acrivos. The shear-induced migration of particles in concentrated suspensions. J. Fluid Mech., 181: 415–439, 1987.
M.B. Mackaplow and E.S.G. Shaqfeh. A numerical study of the sedimentation of fiber suspensions. J. Fluid Mech., 376: 149–182, 1998.
B. Metzger and J. E. Butler. Irreversibility and chaos: Role of long range hydrodynamic interactions in sheared suspensions. Phys. Rev. E, 82: 051406, 2010.
B. Metzger, M. Nicolas, and E. Guazzelli. Falling clouds of particles in viscous fluids. J. Fluid Mech., 580: 283–301, 2007.
B. Metzger, P. Pham, and J.E. Butler. Irreversibility and chaos: Role of lubrication interactions in sheared suspensions. Phys. Rev. E, 87: 052304, 2013.
P.R. Nott and J.F. Brady. Pressure-driven flow of suspensions: simulation and theory. J. Fluid Mech., 275: 157–199, 1994.
A. Oberbeck. Ueber stationiire Fliissigkeitsbewegungen mit Beriicksichtigung der inneren Reibung. J. reine angew. Math., 81: 62–80, 1876.
J. Park and J.E. Butler. Inhomogeneous distribution of a rigid fibre undergoing rectilinear flow between parallel walls at high Peclet numbers. J. Fluid Mech., 630: 267–298, 2009.

J. Park, J.M. Bricker, and J.E. Butler. Cross-stream migration in dilute solutions of rigid polymers undergoing rectilinear flow near a wall. Phys. Rev. E, 76: 040801(R), 2007.

J. Park, B. Metzger, E. Guazzelli, and J.E. Butler. A cloud of rigid fibres sedimenting in a viscous fluid. J. Fluid Mech., 648: 351–362, 2010.

P. Pham, B. Metzger, and J.E. Butler. Particle dispersion in sheared suspensions: Crucial role of solid-solid contacts. Phys. Fluids, 27: 051701, 2015.

P. Pham, B. Metzger, and J.E. Butler. Origin of critical strain amplitude in periodically sheared suspensions. Physical Review Fluids, 1: 022201(R), 2016.

D.J. Pine, J.P. Gollub, J.F. Brady, and A.M. Leshansky. Chaos and threshold for irreversibility in sheared suspensions. Nature, 438: 997, 2005.

C. Pozrikidis. *Boundary Integral and Singularity Methods for Linearized Viscous Flow*. Cambridge University Press, 1992.

J. Riseman and J. G. Kirkwood. The intrinsic viscosity, translational and rotatory diffusion constants of rod-like macromolecules in solution. J. Chem. Phys., 18: 512–516, 1950.

R.G. Larson. The Structure and Rheology of Complex Fluids. Oxford University Press, New York, 1999.

W.B. Russel, E.J. Hinch, L.G. Leal, and G. Tieffenbruck. Rods falling near a vertical wall. J. Fluid Mech., 83: 273, 1977.

B. Snook, E. Guazzelli, and J.E. Butler. Vorticity alignment of rigid fibers in an oscillatory shear flow: Role of confinement. Phys. Fluids, 24: 121702, 2012.

B. Snook, L.M. Davidson, J.E. Butler, O. Pouliquen, and E. Guazzelli. Normal stress differences in suspensions of rigid fibres. J. Fluid Mech., 758: 486–507, 2014.

K. Yeo and M.R. Maxey. Numerical simulations of concentrated suspensions of mono-disperse particles in a Poiseuille flow. J. Fluid Mech., 682: 491–518, 2011.

The manufacturer's authorised representative in the EU is Springer Nature Customer Service Centre GmbH, Europaplatz 3, 69115 Heidelberg, Germany. If you have any concerns regarding our products, please contact ProductSafety@springernature.com

Printed and bound by CPI Group (UK) Ltd, Croydon, CR0 4YY
30/09/2026
02225258-0008